FIFTH EDITION

Atlas of Skeletal Muscles

Robert J. Stone
Suffolk County Community College

Judith A. Stone
Suffolk County Community College

Boston   Burr Ridge, IL   Dubuque, IA   Madison, WI   New York   San Francisco   St. Louis
Bangkok   Bogotá   Caracas   Kuala Lumpur   Lisbon   London   Madrid   Mexico City
Milan   Montreal   New Delhi   Santiago   Seoul   Singapore   Sydney   Taipei   Toronto

The **McGraw·Hill** Companies

ATLAS OF SKELETAL MUSCLES, FIFTH EDITION
International Edition 2006

Published by McGraw-Hill, a business unit of The McGraw-Hill Companies, Inc. 1221 Avenue of the Americas, New York, NY 10020. Copyright © 2006, 2003, 2000, 1997 by The McGraw-Hill Companies, Inc. All rights reserved. No part of this publication may be reproduced or distributed in any form or by any means, or stored in a database or retrieval system, without the prior written consent of The McGraw-Hill Companies, Inc., including, but not limited to, in any network or other electronic storage or transmission, or broadcast for distance learning.
Some ancillaries, including electronic and print components, may not be available to customers outside the United States.

10 09 08 07 06 05 04 03 02 01
20 09 08 07 06 05
CTF   SLP

Library of Congress Control Number: 2004019992

When ordering this title, use ISBN 007-124479-4

Printed in Singapore

612.74 STO

www.mhhe.com

Dedication

To Karen, Andrew, and Laura, helping us more and more as the years go by.

Contents

CHAPTER SEVEN
Muscles of the Forearm and Hand 129

CHAPTER EIGHT
Muscles of the Hip and Thigh 163

Preface

This book is a study guide and reference for the anatomy and actions of human skeletal muscles. It is designed for use by students of anatomy, physical education, and health-related fields. It also serves as a compact reference for the practicing professional.

The first chapter presents photographic illustrations of the major features of the skeleton. These photos have been selectively enhanced and combined to emphasize important features. They are thus a hybrid between drawings and unretouched photographs. The philosophy embraced is that the teacher function as a lens to focus attention on selected facts and observations. A master numbering system is used so that each structure is labeled with the same number in all drawings.

The second chapter describes through illustration and description the various movements of the body.

In chapters 3 through 9, the origin, insertion, action, and innervation of the skeletal muscles are described, and each muscle is presented on a separate page with a line drawing.

The spinal cord levels of the nerve fibers that innervate each muscle are included in parentheses after the name of each nerve.

Labeled drawings of major muscle groups are presented throughout chapters 3 through 9. Notes and relationships among muscles have been included on many pages.

The drawings include the following important features:

1. Bones and cartilage containing muscle attachments are shaded.
2. Adjacent structures are shown.
3. Muscle fibers are drawn by direction.
4. Muscle fibers are shown on the undersurface of bone and cartilage as dashed lines.
5. Tendons and aponeuroses are shown.
6. Labeled muscle groups are included.

These features aid in visual orientation and understanding of the action of the muscles. We have noticed that many students find it useful to color the illustrations.

Notes have been included on many pages to show how muscles are used. Relationships among many of the muscles have also been indicated where appropriate.

Some users of previous editions have advised that some of the smaller muscles should be enlarged and shown with less skeletal background. We have purposely standardized the skeletal views to allow an appreciation of the relative sizes and positions of the muscles. Since skeletal muscles, at the gross level, are relatively simple anatomical structures, very little additional information would be included by enlargement, and many comparative relationships would be lost.

Since our primary objective is to describe the muscles moving the skeleton, we have not included the muscles of the peritoneum, eye, tympanic cavity, tongue, larynx, pharynx, or palate.

We extend our appreciation to Mr. George Boykin, who was for many years the jolly proprietor of the gross anatomy laboratories at the State University of New York at Stony Brook, for his help and encouragement. We also thank Mr. Vincent Verdisco for his technical advice and the many students who have offered valuable suggestions over the years.

We would also like to thank the many reviewers who have made helpful suggestions for improving past editions of this atlas, as well as Wendy D. Bircher, San Juan College; Rosemary Davenport, Gulf Coast Community College; David Hammerman, Long Island University; Andrew Hufford, Butte College; Linda A. Winkler, University of Pittsburgh for their input on the fifth edition.

Robert J. Stone

Judith A. Stone

CHAPTER ONE

The Skeleton

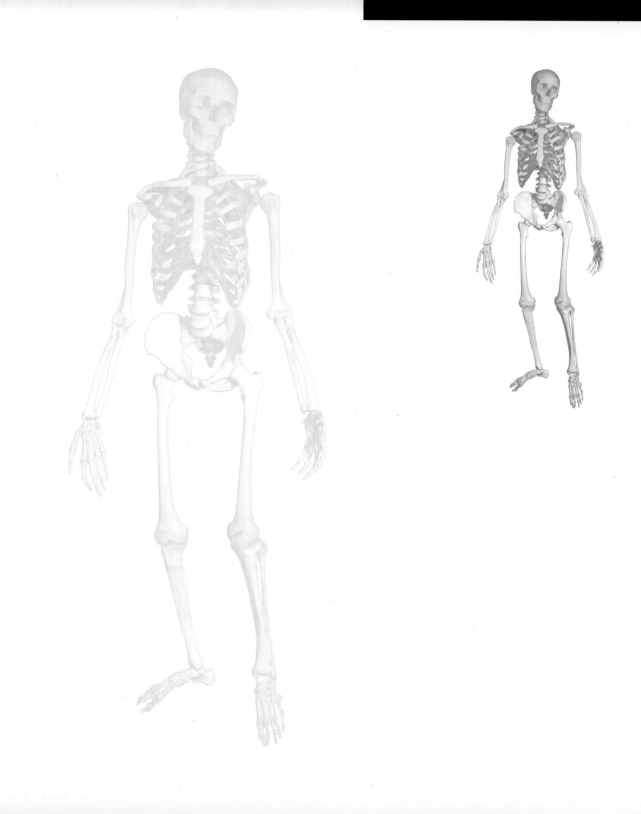

SKULL—ANTERIOR VIEW

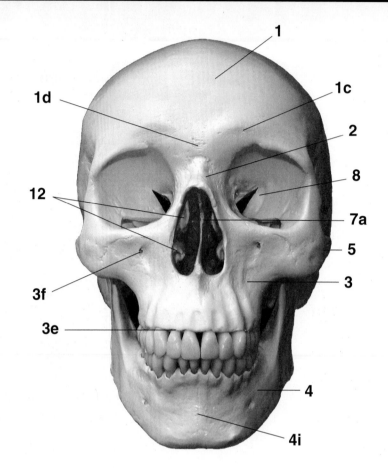

1. Frontal bone
1c. Superciliary arch
1d. Glabella
2. Nasal bone
3. Maxilla
3e. Alveolar border
3f. Infraorbital foramen
3g. Incisive foramen
3h. Palatine process
4. Mandible
4i. Symphysis
5. Zygomatic bone
7a. Nasal septum
8. Sphenoid
8b. Lateral pterygoid plate

8d. Medial pterygoid plate
8e. Foramen ovale
9. Temporal bone
9b. Mastoid process (temporal bone)
9e. Carotid canal
11a. Superior nuchal line (occipital bone)
11b. Inferior nuchal line (occipital bone)
11c. External occipital protuberance (occipital bone)
11d. Occipital condyle (occipital bone)
11e. External occipital crest (occipital bone)
11f. Jugular foramen
11g. Foramen magnum
12. Turbinates
13. Palatine bone
14. Vomer

SKULL—INFERIOR (BASAL) VIEW

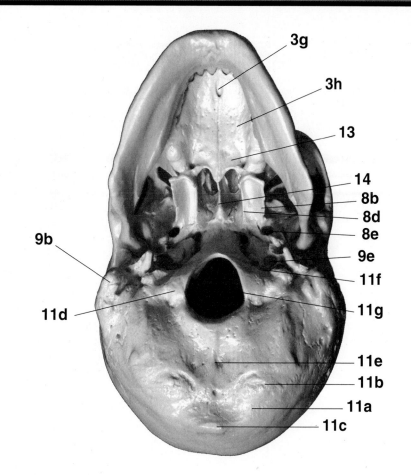

SKULL—LATERAL VIEW

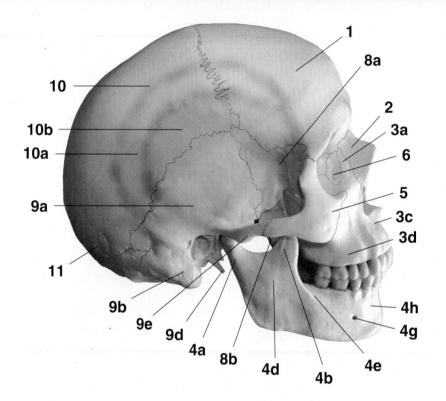

1. Frontal bone
2. Nasal bone
3. Maxilla
3a. Frontal process
3c. Incisive process
3d. Canine fossa
4. Mandible
4a. Neck of condyle
4b. Coronoid process
4c. Angle
4d. Ramus
4e. Oblique line
4g. Mental foramen
4h. Incisive fossa
5. Zygomatic bone
6. Lacrimal bone
8a. Greater wing of sphenoid bone
8b. Lateral pterygoid plate

9. Temporal bone
9a. Temporal fossa (squamous part of temporal bone)
9b. Mastoid process
9d. Styloid process
9e. Zygomatic process
10. Parietal bone
10a. Superior temporal line
10b. Inferior temporal line
11. Occipital bone

Note: The zygomatic arch is formed by the zygomatic process of the temporal bone meeting the zygomatic bone.

VERTEBRAL COLUMN—LATERAL VIEW

C. Cervical vertebrae
C1. Atlas
C2. Axis
C7. Seventh cervical vertebra
T. Thoracic vertebrae
T1. First thoracic vertebra
T12. Twelfth thoracic vertebra
L. Lumbar vertebrae
L1. First lumbar vertebra
L5. Fifth lumbar vertebra
S. Sacrum
Co. Coccyx

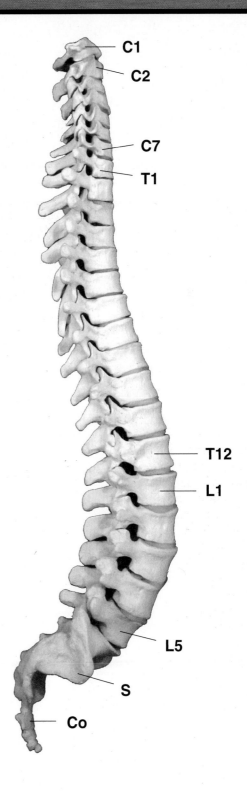

ATLAS AND AXIS

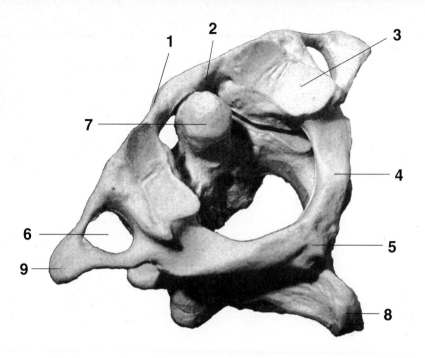

Atlas

1. Anterior tubercle
2. Anterior arch
3. Superior articular facet
4. Posterior arch
5. Posterior tubercle
6. Transverse foramen

Axis

7. Dens
8. Spinous process
9. Transverse process

THORACIC VERTEBRAE—LATERAL VIEW

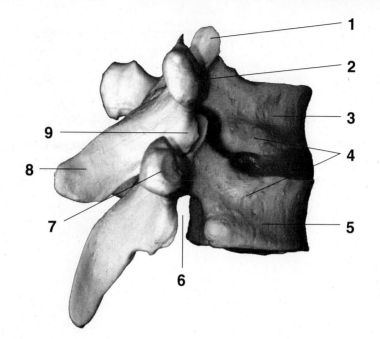

1. Superior articular process
2. Transverse process
3. Body
4. Demifacets (for ribs)
5. Disk
6. Inferior vertebral notch
7. Facet (for rib tubercle)
8. Spinous process
9. Inferior articular process

LUMBAR VERTEBRA—SUPERIOR VIEW

1. Spinous process
2. Mammillary process
3. Transverse process
4. Lamina
5. Vertebral foramen
6. Pedicle
7. Body (centrum)

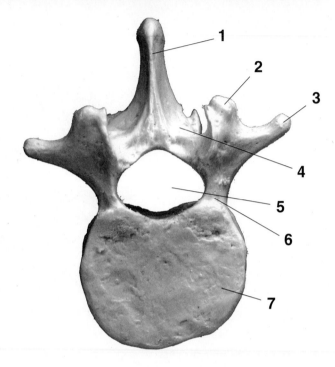

SACRUM—POSTERIOR VIEW

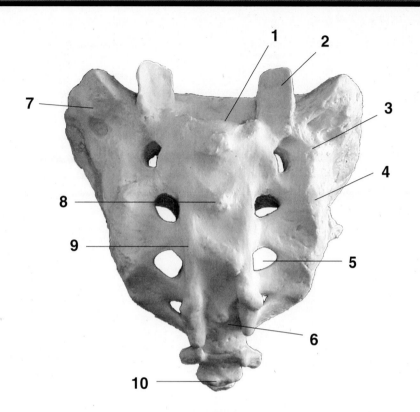

1. Sacral canal
2. Superior articular facet
3. Sacral tuberosity
4. Lateral crest
5. Posterior sacral foramen
6. Sacral hiatus (opening of sacral canal)
7. Auricular surface
8. Median crest
9. Intermediate crest
10. Coccyx

SACRUM—PELVIC VIEW

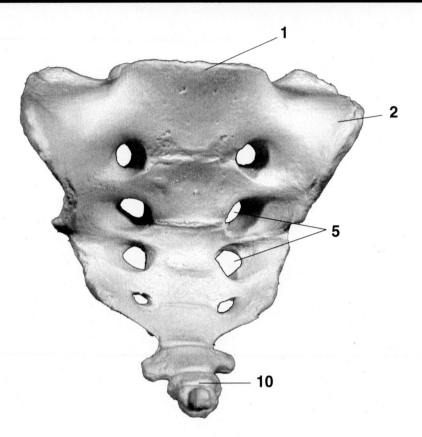

1. Promontory
2. Ala
5. Sacral foramina
10. Coccyx

STERNUM AND CLAVICLE WITH SCAPULA

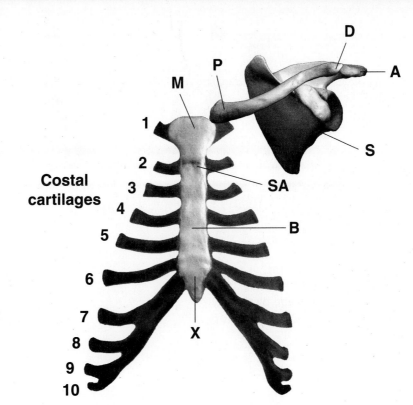

Clavicle
P = proximal end
D = distal end
Scapula (S)
A = acromion
Sternum
M = manubrium
SA = sternal angle
B = body
X = xiphoid process

The proximal (sternal) end of the clavicle forms the sternoclavicular joint with the manubrium of the sternum.

The distal (scapular) end forms the acromioclavicular joint with the acromion of the scapula. This is the only bony articulation of the upper limb with the torso.

The second rib cartilage articulates at the sternal angle between the manubrium and the body of the sternum.

The cartilages of ribs 7-10 are fused to form the costal arch.

HYOID BONE—SUPERIOR VIEW

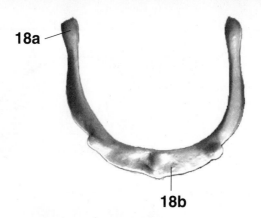

18a

18b

18. Hyoid bone
18a. Greater horn
18b. Body

The greater horn of the hyoid bone attaches to the styloid processes of the temporal bones by the styloid ligaments. It attaches to the thyroid cartilage by the thyrohyoid ligament and supports the upper respiratory tract.

RIB ARTICULATIONS

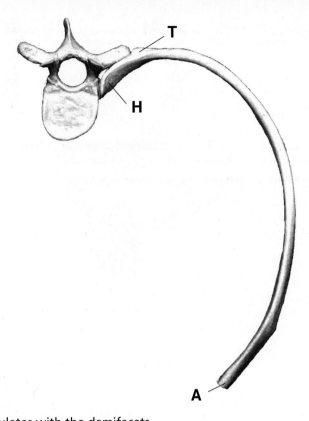

The head (H) of a rib articulates with the demifacets of two adjacent vertebrae. The tubercle (T) joins the facet of the transverse process of the upper vertebra. The anterior end (A) meets the costal cartilage which then joins the sternum.

SCAPULA AND HUMERUS—ANTERIOR VIEW

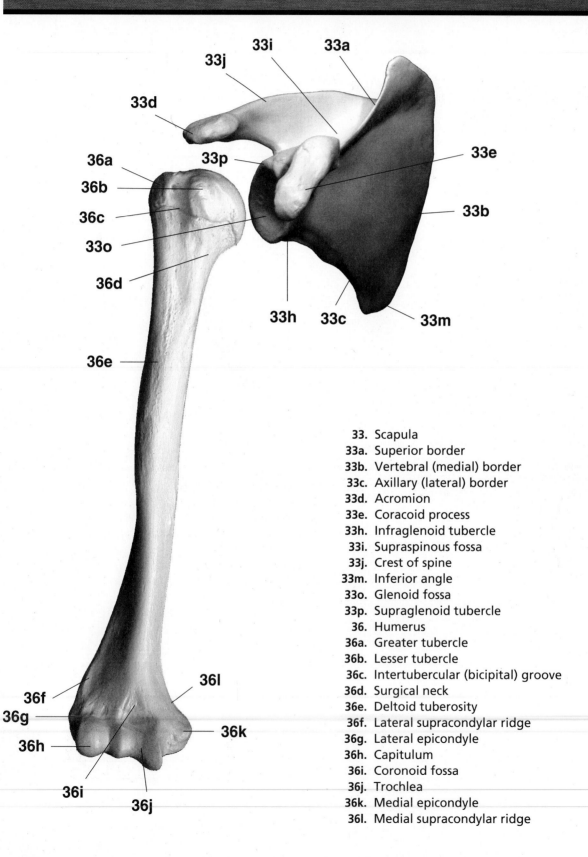

33. Scapula
33a. Superior border
33b. Vertebral (medial) border
33c. Axillary (lateral) border
33d. Acromion
33e. Coracoid process
33h. Infraglenoid tubercle
33i. Supraspinous fossa
33j. Crest of spine
33m. Inferior angle
33o. Glenoid fossa
33p. Supraglenoid tubercle
36. Humerus
36a. Greater tubercle
36b. Lesser tubercle
36c. Intertubercular (bicipital) groove
36d. Surgical neck
36e. Deltoid tuberosity
36f. Lateral supracondylar ridge
36g. Lateral epicondyle
36h. Capitulum
36i. Coronoid fossa
36j. Trochlea
36k. Medial epicondyle
36l. Medial supracondylar ridge

SCAPULA AND HUMERUS—POSTERIOR VIEW

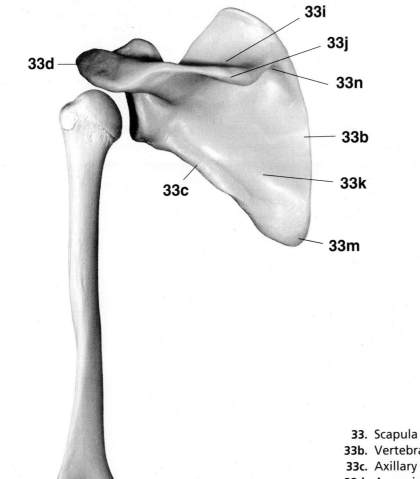

33. Scapula
33b. Vertebral (medial) border
33c. Axillary (lateral) border
33d. Acromion
33i. Supraspinous fossa
33j. Crest of spine
33k. Infraspinous fossa
33m. Inferior angle
33n. Root of spine
36. Humerus
36g. Lateral epicondyle
36k. Medial epicondyle
36m. Olecranon fossa

ELBOW—ANTERIOR VIEW

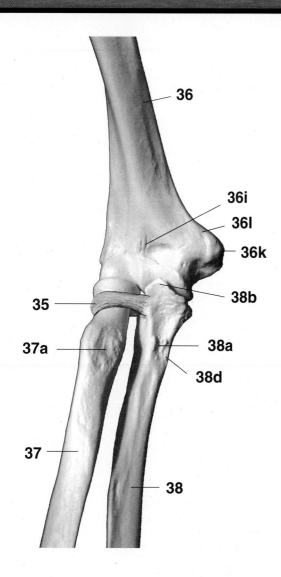

35. Annular ligament
36. Humerus
36i. Coronoid fossa
36k. Medial epicondyle
36l. Medial supracondylar ridge
37. Radius
37a. Radial tuberosity
38. Ulna
38a. Ulnar tuberosity
38b. Coronoid process
38d. Supinator crest

FOREARM—POSTERIOR VIEW

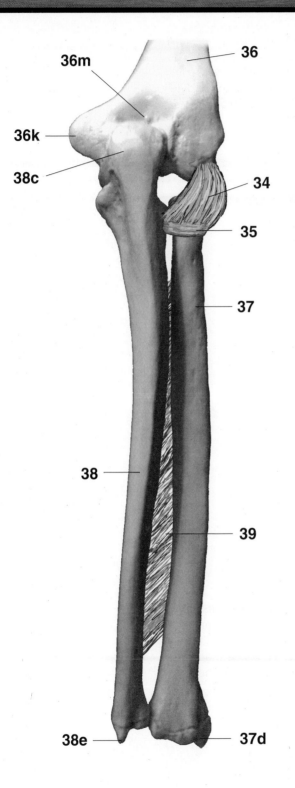

34. Radial (lateral) collateral ligament
35. Annular ligament
36. Humerus
36k. Medial epicondyle
36m. Olecranon fossa
37. Radius
37d. Styloid process
38. Ulna
38c. Olecranon
38e. Styloid process
39. Interosseous membrane

HAND—PALMAR VIEW

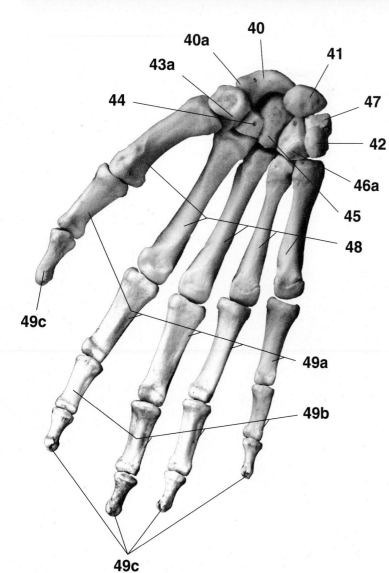

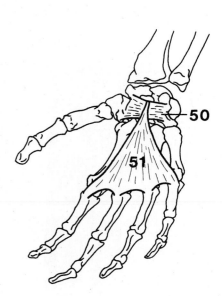

40. Scaphoid
40a. Tubercle of scaphoid
41. Lunate
42. Pisiform
43. Trapezium
43a. Tubercle of trapezium
44. Trapezoid
45. Capitate
46. Hamate
46a. Hook of hamate
47. Triquetrum
48. Metacarpals
49. Phalanges

49a. Proximal (first) phalanges
49b. Middle (second) phalanges
49c. Distal (third) phalanges
50. Flexor retinaculum
51. Palmar aponeurosis

HAND—DORSAL VIEW

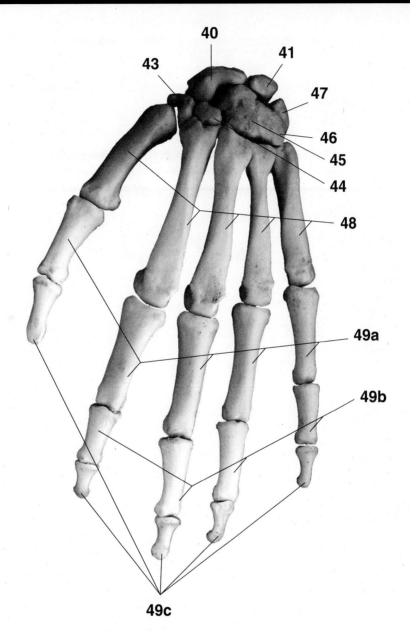

40. Scaphoid
41. Lunate
43. Trapezium
44. Trapezoid
45. Capitate
46. Hamate
47. Triquetrum
48. Metacarpals
49. Phalanges

49a. Proximal (first) phalanges
49b. Middle (second) phalanges
49c. Distal (third) phalanges

PELVIS—ANTERIOR VIEW

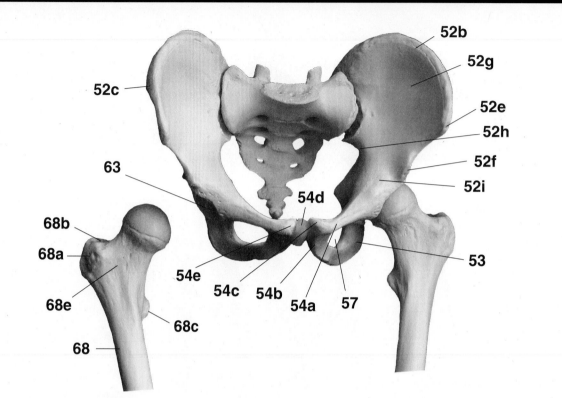

52. Ilium
52b. Iliac crest
52c. Iliac tubercle
52e. Anterior superior iliac spine
52f. Anterior inferior iliac spine
52g. Iliac fossa
52h. Arcuate line
52i. Iliopectineal eminence
53. Ischium
54. Pubis
54a. Superior ramus
54b. Inferior ramus
54c. Pubic crest
54d. Pubic symphysis
54e. Pubic tubercle
57. Obturator foramen
63. Acetabulum
68. Femur
68a. Greater trochanter
68b. Trochanteric fossa
68c. Lesser trochanter
68e. Intertrochanteric line

PELVIS—THREE QUARTER POSTERIOR VIEW

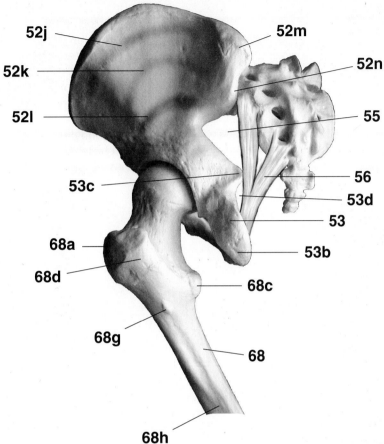

52j
52k
52l
53c
68a
68d
68g
68h

52m
52n
55
56
53d
53
53b
68c
68

52. Ilium
52j. Posterior gluteal line
52k. Middle gluteal line
52l. Inferior gluteal line
52m. Posterior superior iliac spine
52n. Posterior inferior iliac spine
53. Ischium
53b. Ischial tuberosity
53c. Ischial spine
53d. Lesser sciatic notch
55. Greater sciatic notch
56. Sacrotuberous ligament
68. Femur
68a. Greater trochanter
68c. Lesser trochanter
68d. Intertrochanteric crest
68g. Gluteal tuberosity
68h. Linea aspera

FEMUR—POSTERIOR VIEW

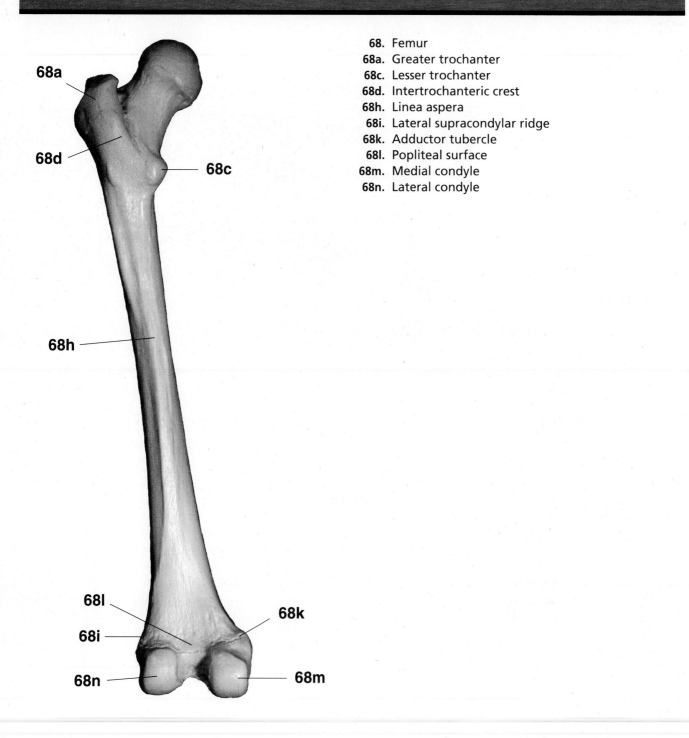

68. Femur
68a. Greater trochanter
68c. Lesser trochanter
68d. Intertrochanteric crest
68h. Linea aspera
68i. Lateral supracondylar ridge
68k. Adductor tubercle
68l. Popliteal surface
68m. Medial condyle
68n. Lateral condyle

KNEE—ANTERIOR VIEW

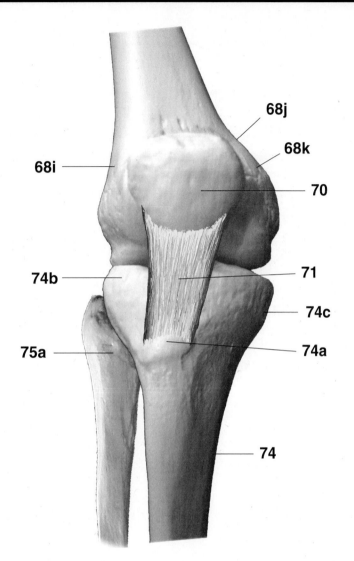

68. Femur
68i. Lateral supracondylar ridge
68j. Medial supracondylar ridge
68k. Adductor tubercle
70. Patella
71. Patellar ligament
74. Tibia
74a. Tibial tuberosity
74b. Lateral condyle
74c. Medial condyle
75. Fibula
75a. Head of fibula

Note: The quadriceps tendon continues through the
patella and becomes the patellar ligament.

ANKLE AND FOOT—ANTEROLATERAL VIEW

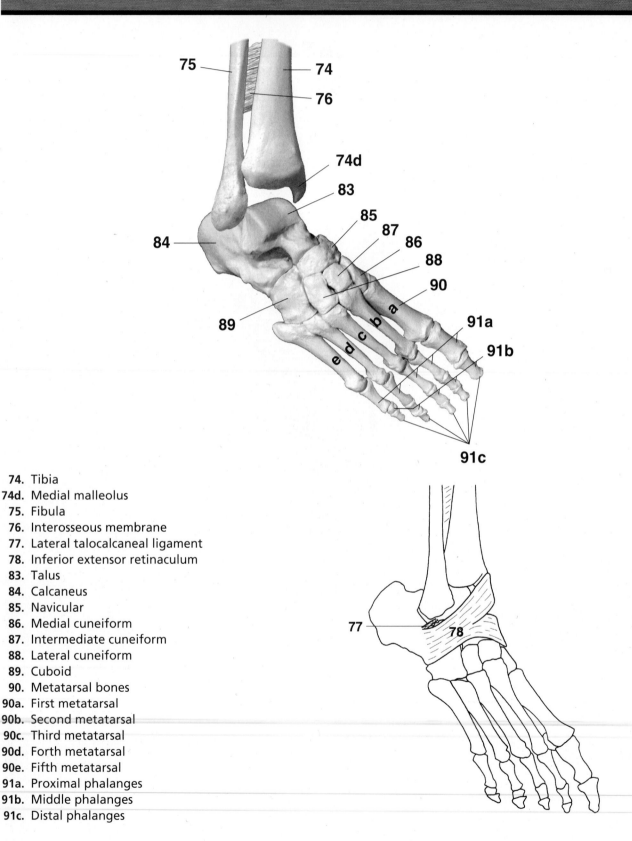

74. Tibia
74d. Medial malleolus
75. Fibula
76. Interosseous membrane
77. Lateral talocalcaneal ligament
78. Inferior extensor retinaculum
83. Talus
84. Calcaneus
85. Navicular
86. Medial cuneiform
87. Intermediate cuneiform
88. Lateral cuneiform
89. Cuboid
90. Metatarsal bones
90a. First metatarsal
90b. Second metatarsal
90c. Third metatarsal
90d. Forth metatarsal
90e. Fifth metatarsal
91a. Proximal phalanges
91b. Middle phalanges
91c. Distal phalanges

FOOT—PLANTAR VIEW

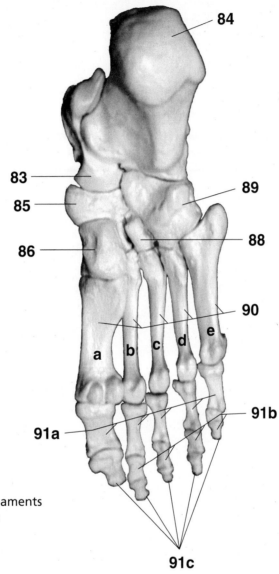

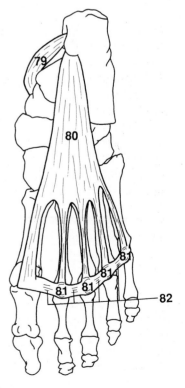

79. Flexor retinaculum
80. Plantar aponeurosis
81. Plantar metatarsophalangeal ligaments
82. Transverse metatarsal ligaments
83. Talus
84. Calcaneus
85. Navicular
86. Medial cuneiform
88. Lateral cuneiform
89. Cuboid
90. Metatarsal bones
90a. First metatarsal
90b. Second metatarsal
90c. Third metatarsal
90d. Forth metatarsal
90e. Fifth metatarsal
91a. Proximal phalanges
91b. Middle phalanges
91c. Distal phalanges

CHAPTER TWO

Movements of the Body

Anatomical position—A subject in the anatomical position is standing erect with the head, eyes, and toes facing forward and the arms hanging straight at the sides with the palms of the hands facing forward.

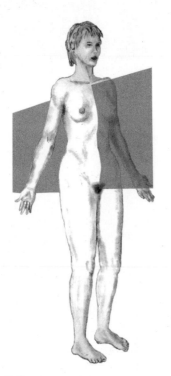

FIGURE 2.1
Median or midsagittal plane—Passes vertically through the body from anterior (front) to posterior (back). It divides the body into right and left sides. Other sagittal planes are parallel to this plane.

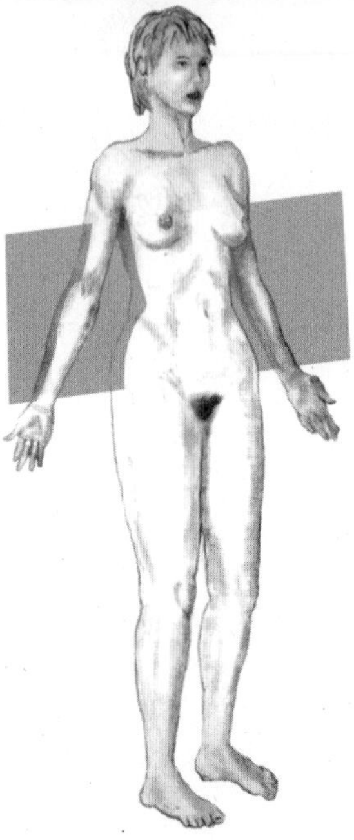

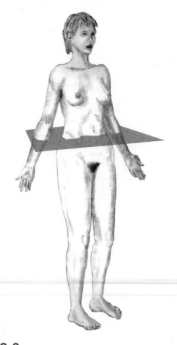

FIGURE 2.2
Coronal (frontal) planes—Pass vertically through the body from side to side. They divide the body from front to back.

FIGURE 2.3
Transverse planes (cross sections)—Pass horizontally through the body parallel to the ground.

F I G U R E 2.4
Flexion—The left arm, forearm, and right thigh are drawn forward in sagittal planes. The right knee is also flexed.
Extension—The left thigh and knee are extended.
Hyperextension—The right arm is hyperextended at the shoulder.

F I G U R E 2.5
Lateral flexion—The torso (or head) bends laterally in the coronal plane.

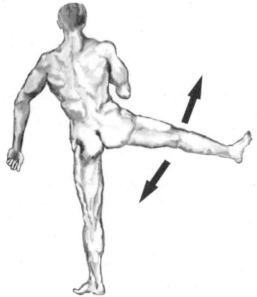

F I G U R E 2.6
Abduction—The right leg moves laterally in the coronal plane.
Adduction—The leg is returned medially in the coronal plane.

Note: If the foot is fixed, this motion results in tilting the pelvis upward on the opposite side.

F I G U R E 2.7
Medial rotation—The anterior of the arm (or thigh) is moved toward the median plane.
Lateral rotation—The anterior of the arm (or thigh) is moved away from the median plane.

MOVEMENTS OF THE SCAPULA

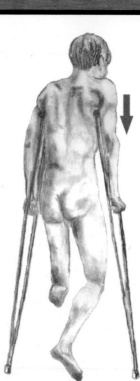

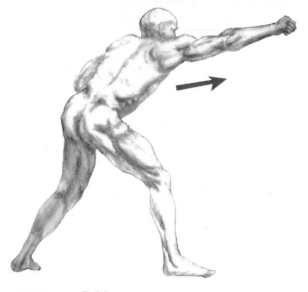

FIGURE 2.8
Elevation—The right scapula of this figure is drawn superiorly against the resistance of the rock.

FIGURE 2.9
Depression—The right scapula of this figure is pushing the arm inferiorly.

FIGURE 2.10
Protraction—The scapula pushes the arm forward in a sagittal plane.

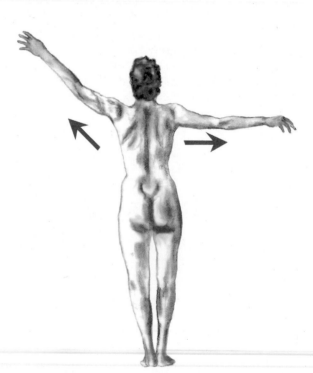

FIGURE 2.11
Retraction—The scapula is pulled back from protraction in a sagittal plane. Since the scapula slides around the ribs toward the median plane, it becomes adduction.

FIGURE 2.12
Rotation—For abduction of the arm to continue above the height of the shoulder, the scapula must rotate on its axis so that the glenoid fossa turns upward.

MOVEMENTS OF THE HAND AND FOREARM

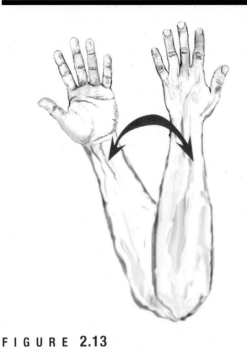

FIGURE 2.13
Pronation—The forearm is rotated away from the anatomical position so that the palm turns medially, then posteriorly. If the forearm is flexed at the elbow, then the palm turns inferiorly.
Supination—The forearm is rotated so that the palm turns anteriorly (or superiorly if the forearm is flexed). Also see figs. 2.19 and 2.20.

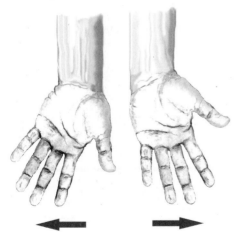

FIGURE 2.15
Radial flexion (abduction)—The hand, at the wrist, is drawn away from the body in a coronal plane.
Ulnar flexion (adduction)—The hand, at the wrist, is drawn toward the body in a coronal plane.

FIGURE 2.14
Abduction—The fingers are moved away from the midline of the hand.

FIGURE 2.16
Adduction—The fingers are moved toward the midline of the hand.

FIGURE 2.17
Opposition—The thumb is rotated so its anterior pad can touch the anterior pads of the four fingers.

MOVEMENTS OF THE FOOT

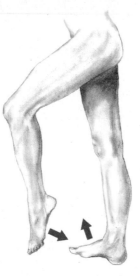

FIGURE 2.18
Dorsiflexion—The ankle flexes, moving the foot superiorly.
Plantar Flexion—The ankle extends, moving the foot inferiorly.

FIGURE 2.19
Eversion—The front of the foot moves laterally away from the midline (abduction), and the sole turns outward. This is also called pronation.

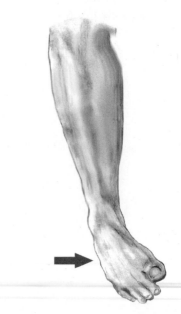
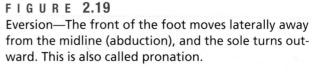

FIGURE 2.20
Inversion—The front of the foot moves medially toward the midline (adduction), and the sole turns inward. This is also called supination.

CHAPTER THREE

Muscles of the Face and Head

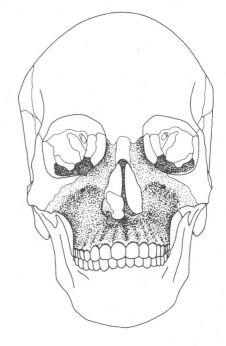

EPICRANIUS

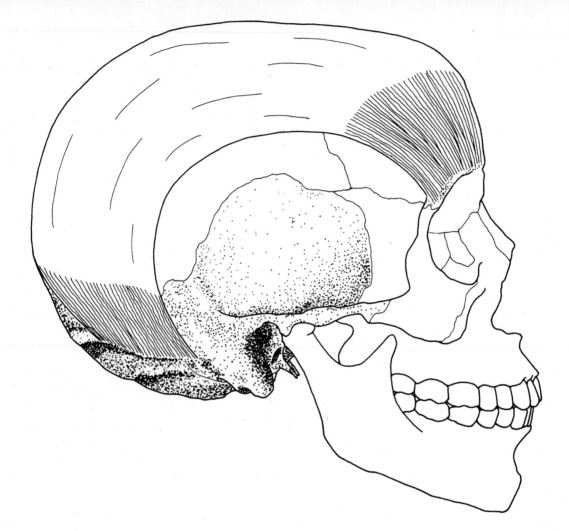

Skull—lateral view

Occipital belly (*occipitalis*)

■ **Origin**
Lateral two-thirds of superior nuchal line of occipital bone, mastoid process of temporal bone

■ **Insertion**
Galea aponeurotica (an intermediate tendon leading to frontal belly)

■ **Action**
Draws back scalp, aids frontal belly to wrinkle forehead and raise eyebrows

■ **Nerve**
Posterior auricular branch of facial nerve

Frontal belly (*frontalis*)

■ **Origin**
Galea aponeurotica

■ **Insertion**
Fascia of facial muscles and skin above nose and eyes

■ **Action**
Draws back scalp, wrinkles forehead, raises eyebrows

■ **Nerve**
Temporal branches of facial nerve

TEMPOROPARIETALIS

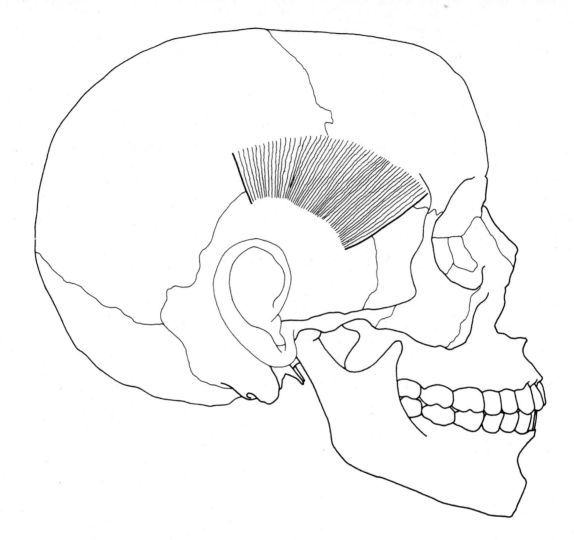

Skull—lateral view

■ **Origin**
Fascia over ear

■ **Insertion**
Lateral border of galea aponeurotica

■ **Action**
Raises ears, tightens scalp

■ **Nerve**
Temporal branch of facial nerve

AURICULARIS ANTERIOR, SUPERIOR, POSTERIOR

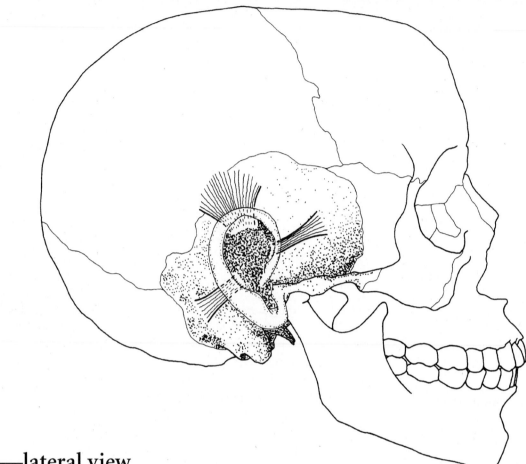

Skull—lateral view

Auricularis anterior

■ **Origin**
Fascia in temporal region

■ **Insertion**
Anterior to helix of ear

■ **Action**
Draws ear forward in some individuals, moves scalp*

■ **Nerve**
Temporal branch of facial nerve

Auricularis superior

■ **Origin**
Fascia in temporal region

■ **Insertion**
Superior part of ear

■ **Action**
Draws ear upward in some individuals, moves scalp*

■ **Nerve**
Temporal branch of facial nerve

Auricularis posterior

■ **Origin**
Mastoid area of temporal bone

■ **Insertion**
Posterior part of ear

■ **Action**
Draws ear upward in some individuals*

■ **Nerve**
Posterior auricular branch of facial nerve

*This muscle is nonfunctional in most people.

ORBICULARIS OCULI

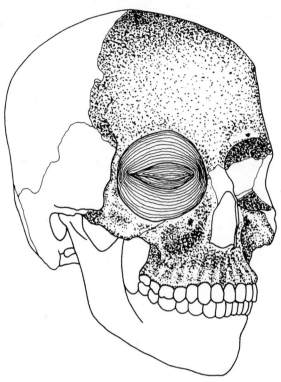

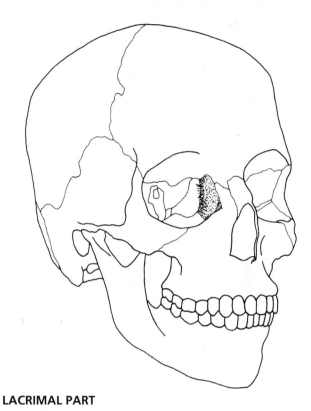

ORBITAL AND PALPEBRAL PARTS

LACRIMAL PART

Skull—three-quarter anterior view

Orbital part

■ **Origin**
Frontal bone, maxilla (medial margin of orbit)

■ **Insertion**
Continues around orbit and returns to origin

■ **Action**
Strong closure of eyelids

■ **Nerve**
Temporal and zygomatic branches of facial nerve

Palpebral part *(in eyelids)*

■ **Origin**
Medial palpebral ligament

■ **Insertion**
Lateral palpebral ligament into zygomatic bone

■ **Action**
Gentle closure of eyelids

■ **Nerve**
Temporal and zygomatic branches of facial nerve

Lacrimal part *(behind medial palpebral ligament and lacrimal sac)*

■ **Origin**
Lacrimal bone

■ **Insertion**
Lateral palpebral raphe

■ **Action**
Draws lacrimal canals onto surface of eye

■ **Nerve**
Temporal and zygomatic branches of facial nerve

LEVATOR PALPEBRAE SUPERIORIS

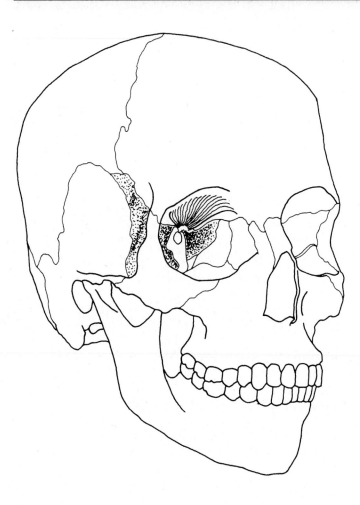

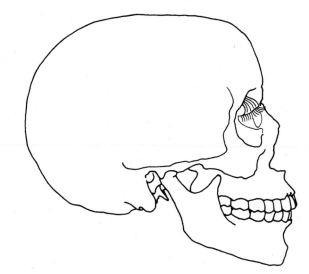

Skull—three-quarter anterior view

■ Origin
Inferior surface of lesser wing of sphenoid

■ Insertion
Skin of upper eyelid

■ Action
Raises upper eyelid

■ Nerve
Oculomotor nerve

Skull—lateral view

CORRUGATOR SUPERCILII

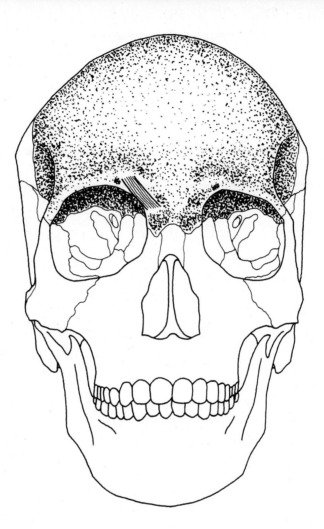

Skull—anterior view

■ Origin
Medial end of superciliary arch

■ Insertion
Deep surface of skin under medial portion of eyebrows

■ Action
Draws eyebrows downward and medially

■ Nerve
Temporal branch of facial nerve

PROCERUS

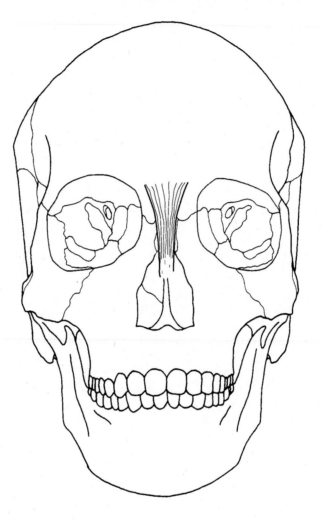

Skull—anterior view

■ **Origin**
Fascia over nasal bone and lateral nasal cartilage

■ **Insertion**
Skin between eyebrows

■ **Action**
Draws down medial part of eyebrows, wrinkles nose

■ **Nerve**
Buccal branches of facial nerve

NASALIS

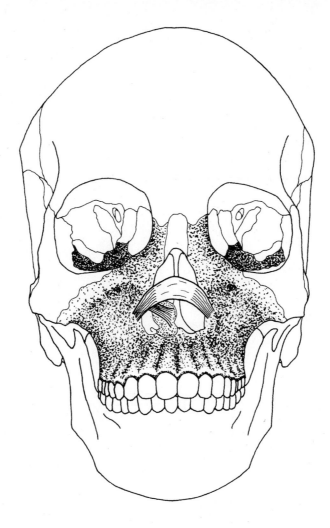

Skull—anterior view

Transverse part

■ **Origin**
Middle of maxilla

■ **Insertion**
Muscle of opposite side over bridge of nose

Alar part

■ **Origin**
Greater alar cartilage, skin on nose

■ **Insertion**
Skin at point of nose

■ **Action**
Both parts maintain opening of external nares during forceful inspiration

■ **Nerve**
Buccal branches of facial nerve

DEPRESSOR SEPTI

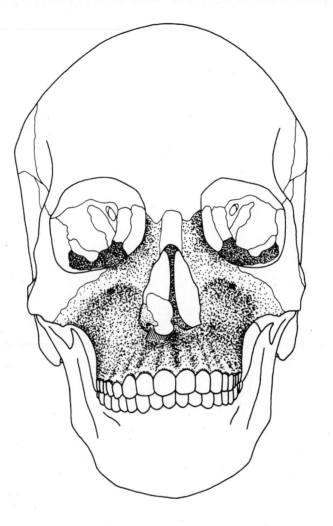

Skull—anterior view

■ **Origin**
Incisive fossa of maxilla

■ **Insertion**
Nasal septum and ala

■ **Action**
Constricts nares

■ **Nerve**
Buccal branches of facial nerve

ORBICULARIS ORIS

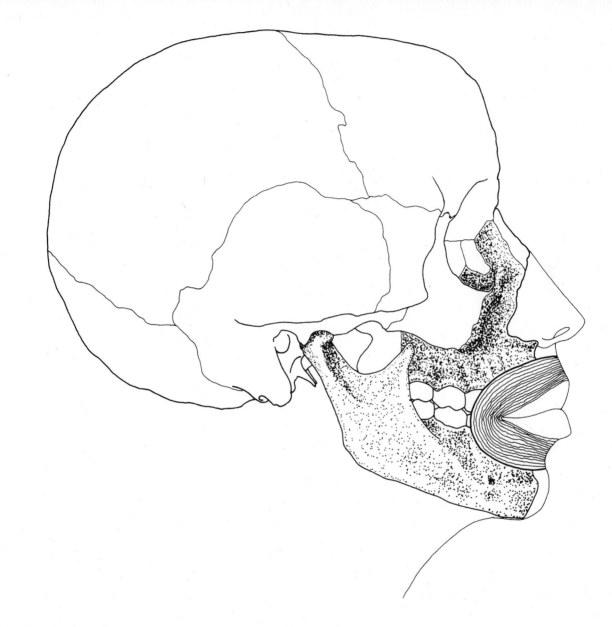

Skull—lateral view

■ Origin
Lateral band—alveolar border of maxilla

Medial band—septum of nose

Inferior portion—lateral to midline of mandible

■ Insertion
Becomes continuous with other muscles at angle of mouth

■ Action
Closure and protrusion of lips

■ Nerve
Buccal and mandibular branches of facial nerve

LEVATOR LABII SUPERIORIS

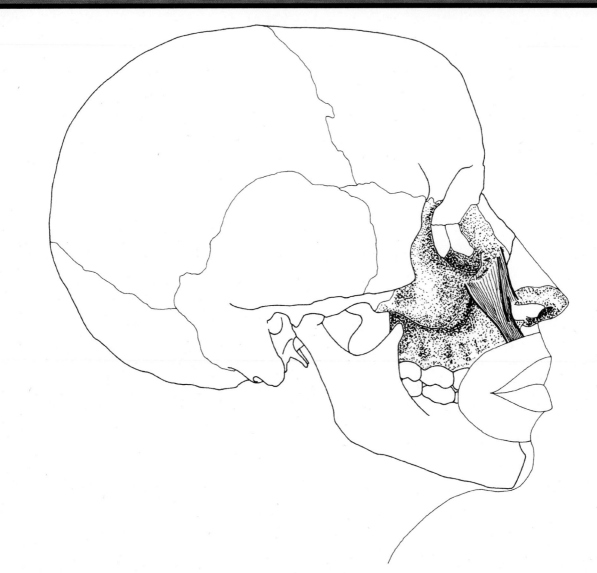

Skull—lateral view

Angular head	Infraorbital head
■ Origin	**■ Origin**
Frontal process of maxilla and zygomatic bone	Lower margin of orbit
■ Insertion	**■ Insertion**
Greater alar cartilage and skin of nose, upper lip	Muscles of upper lip
■ Action	**■ Action**
Elevates upper lip, dilates nares, forms nasolabial furrow	Elevates upper lip
■ Nerve	**■ Nerve**
Buccal branches of facial nerve	Buccal branches of facial nerve

LEVATOR ANGULI ORIS

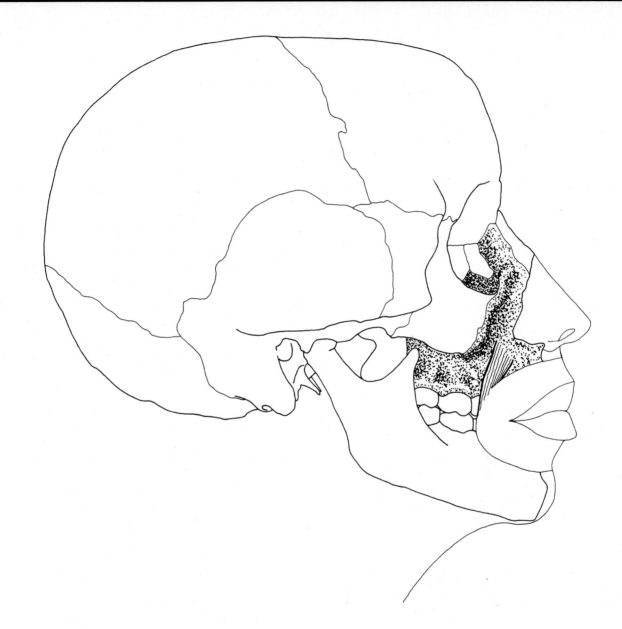

Skull—lateral view

■ **Origin**
Canine fossa of maxilla

■ **Insertion**
Angle of mouth

■ **Action**
Elevates corner (angle) of mouth

■ **Nerve**
Buccal branches of facial nerve

ZYGOMATICUS MAJOR

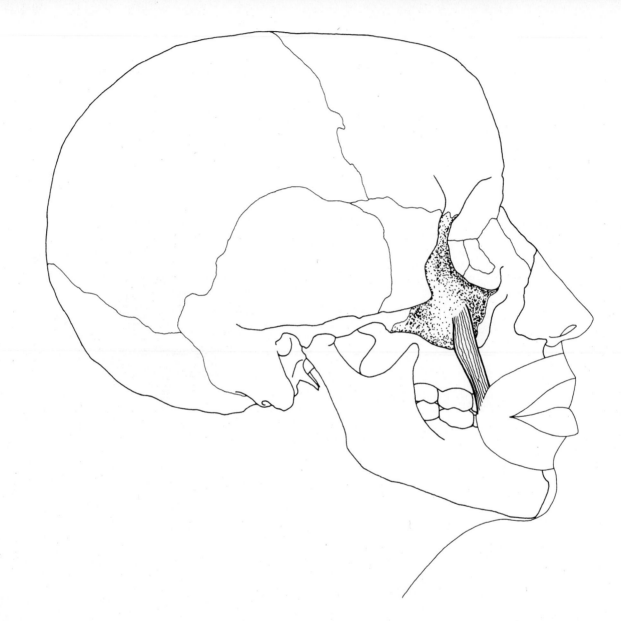

Skull—lateral view

■ **Origin**
Zygomatic bone

■ **Insertion**
Angle of mouth

■ **Action**
Draws angle of mouth upward and backward (laughing)

■ **Nerve**
Buccal branches of facial nerve

ZYGOMATICUS MINOR

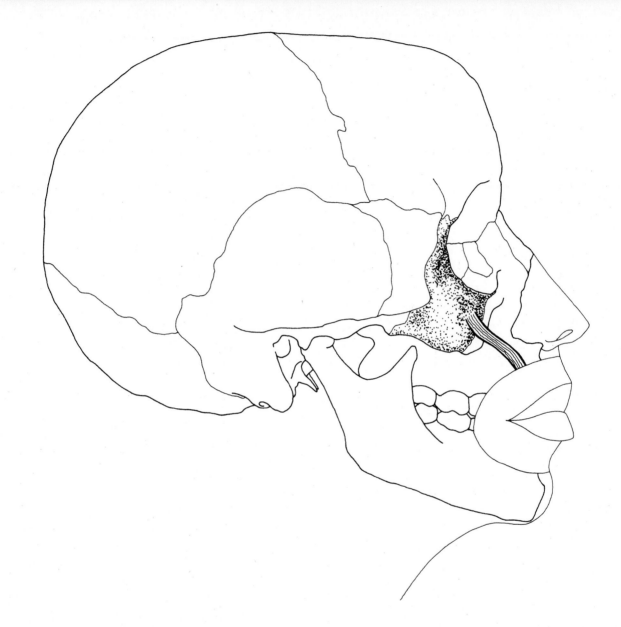

Skull—lateral view

■ Origin
Zygomatic bone

■ Insertion
Upper lip lateral to levator labii superioris

■ Action
Forms nasolabial furrow

■ Nerve
Buccal branches of facial nerve

RISORIUS

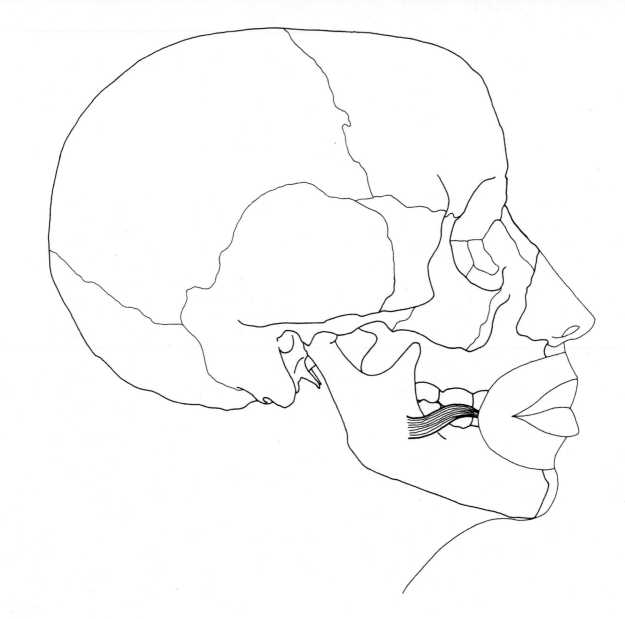

Skull—lateral view

- ■ **Origin**
Fascia over masseter

- ■ **Insertion**
Skin at angle of mouth

- ■ **Action**
Retracts angle of mouth, as in grinning

- ■ **Nerve**
Buccal branches of facial nerve

DEPRESSOR LABII INFERIORIS

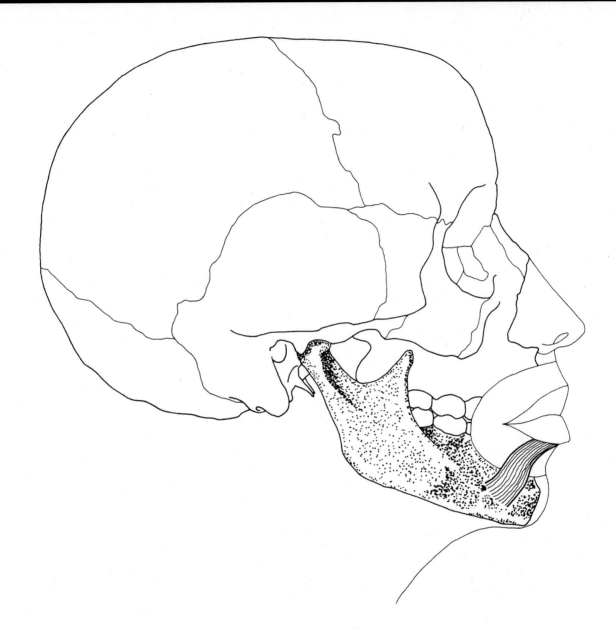

Skull—lateral view

■ Origin
Mandible, between symphysis and mental foramen

■ Insertion
Skin of lower lip

■ Action
Draws lower lip downward and laterally

■ Nerve
Mandibular branch of facial nerve

DEPRESSOR ANGULI ORIS

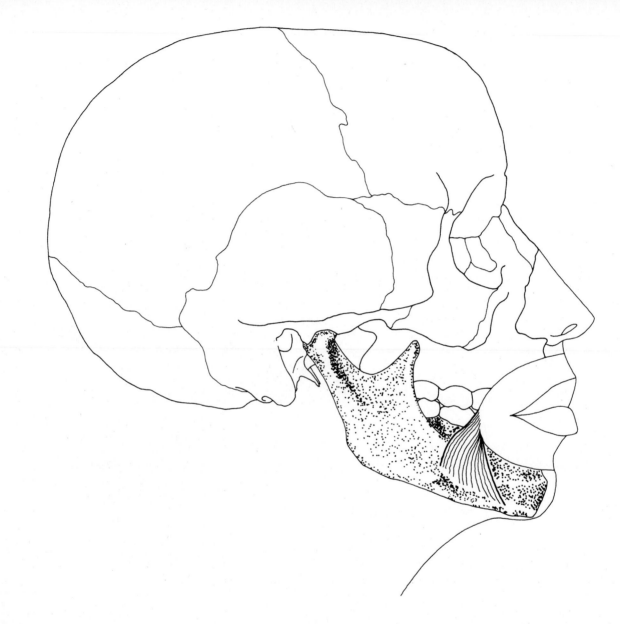

Skull—lateral view

■ **Origin**
Oblique line of the mandible

■ **Insertion**
Angle of the mouth

■ **Action**
Depresses angle of mouth, as in frowning

■ **Nerve**
Mandibular branch of facial nerve

MENTALIS

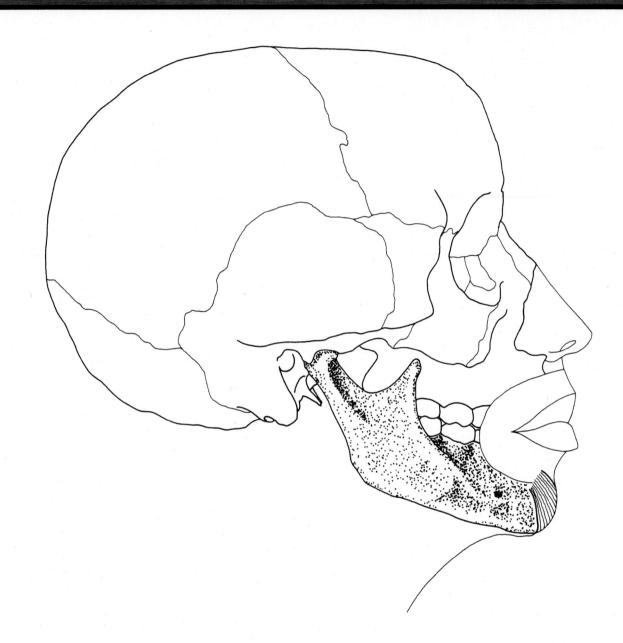

Skull—lateral view

■ **Origin**
Incisive fossa of mandible

■ **Insertion**
Skin of chin

■ **Action**
Raises and protrudes lower lip, wrinkles skin of chin

■ **Nerve**
Mandibular branch of facial nerve

BUCCINATOR

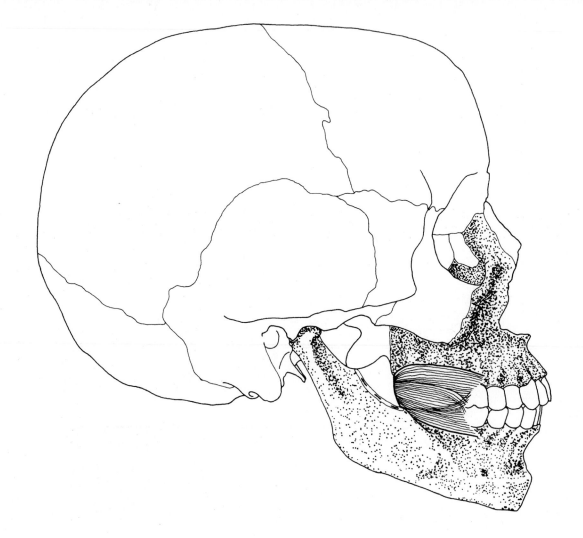

Skull—lateral view

■ Origin
Outer surface of alveolar processes of maxilla and mandible over molars and along pterygomandibular raphe

■ Insertion
Deep part of muscles of lips

■ Action
Compresses cheek

■ Nerve
Buccal branches of facial nerve

TEMPORALIS

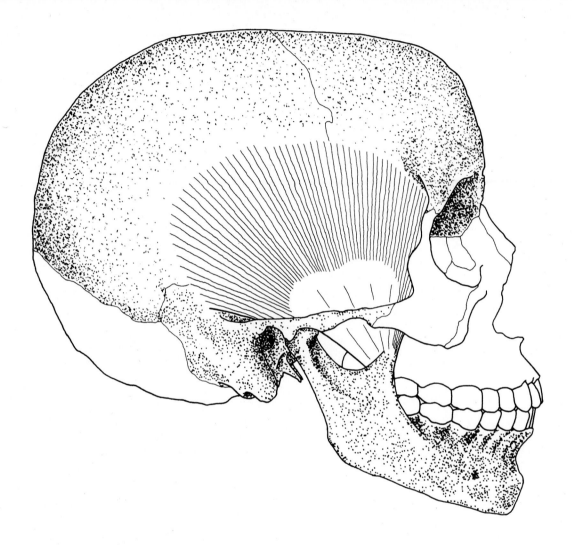

Skull—lateral view

■ Origin
Temporal fossa including frontal, parietal, and temporal bones

■ Insertion
Coronoid process and anterior border of ramus of mandible

■ Action
Closes lower jaw, clenches teeth

■ Nerve
Mandibular division of trigeminal nerve

MASSETER

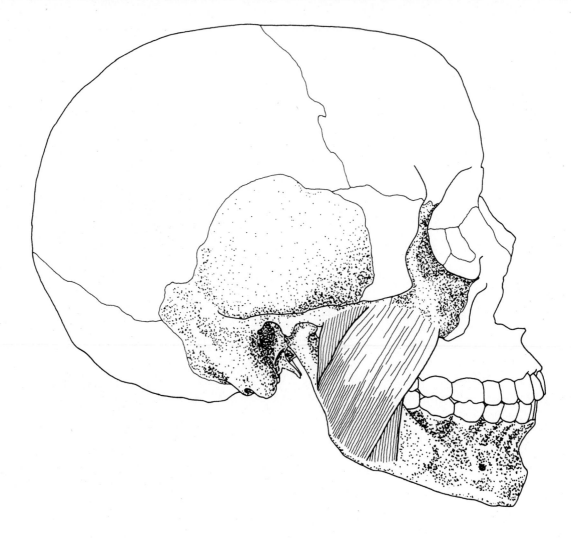

Skull—lateral view

■ Origin
Zygomatic process of maxilla, medial and inferior surfaces of zygomatic arch

■ Insertion
Angle and ramus of mandible, lateral surface of coronoid process of mandible

■ Action
Closes lower jaw, clenches teeth

■ Nerve
Mandibular division of trigeminal nerve

Note: Superficial fibers slightly protract jaw (see lateral pterygoid).

PTERYGOIDEUS MEDIALIS *(Medial Pterygoid)*

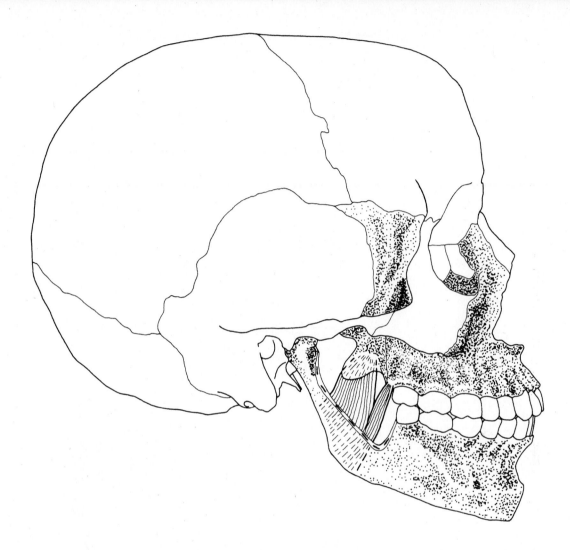

Skull—lateral view
(Part of mandible cut away)

■ Origin
Medial surface of lateral pterygoid plate of sphenoid bone, palatine bone, and tuberosity of maxilla

■ Insertion
Medial surface of ramus and angle of mandible

■ Action
Closes lower jaw, clenches teeth

■ Nerve
Mandibular division of trigeminal nerve

PTERYGOIDEUS LATERALIS *(Lateral Pterygoid)*

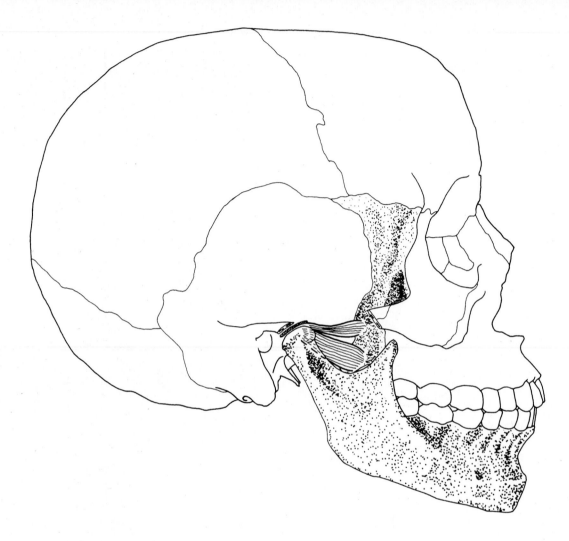

Skull—lateral view

■ Origin

Superior head*—lateral surface of greater wing of sphenoid

Inferior head—lateral surface of lateral pterygoid plate

■ Insertion

Condyle of mandible, temporomandibular joint

■ Action

Opens jaws, protrudes mandible, moves mandible sidewards

■ Nerve

Mandibular division of trigeminal nerve

Note: This sideward movement, aided by superficial fibers of masseter, causes chewing movements.

*Stern calls this a separate muscle: superior pterygoid.

Reference: Stern, J.T.: *Essentials of Gross Anatomy*, F.A. Davis Company, Philadelphia, 1988.

Muscles of
the Neck

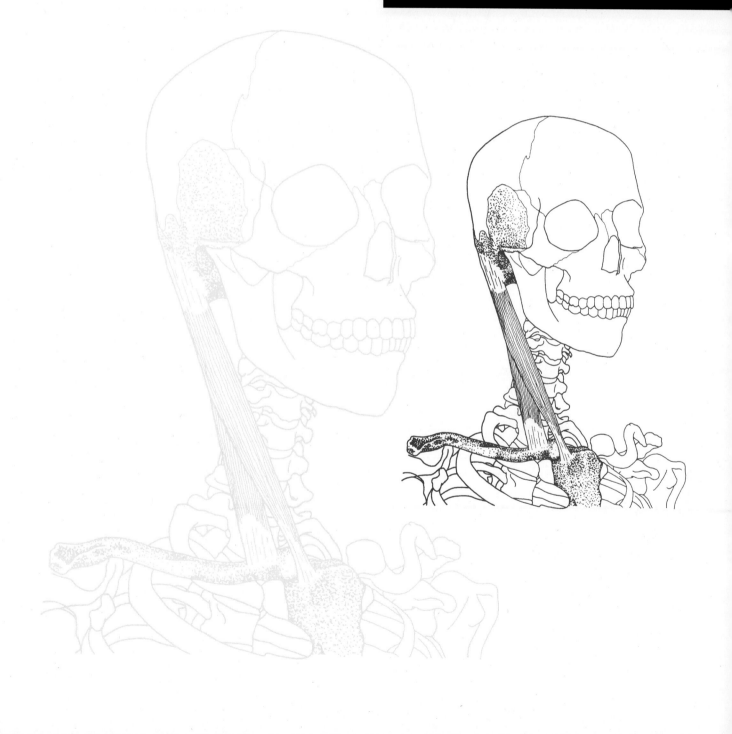

STERNOCLEIDOMASTOID

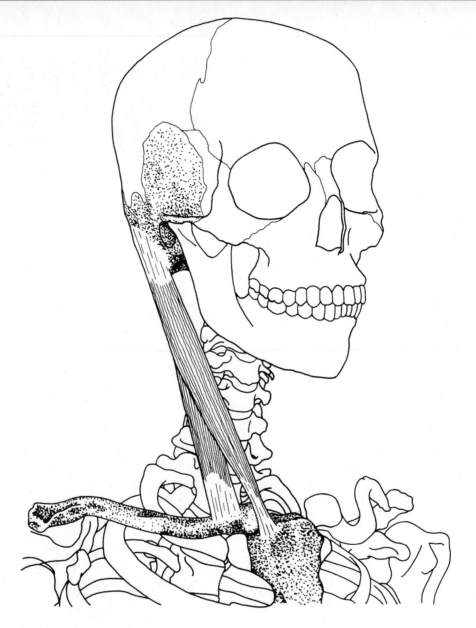

Three-quarter frontal view

■ Origin

Sternal head—manubrium of sternum

Clavicular head—medial part of clavicle

■ Insertion

Mastoid process of temporal bone, lateral half of superior nuchal line of occipital bone

■ Action

One side—bends neck laterally, rotates head to opposite side

Both sides together—flexes neck, draws head ventrally and elevates chin, draws sternum superiorly in deep inspiration

■ Nerve

Spinal part of accessory nerve (C2, C3)

PLATYSMA

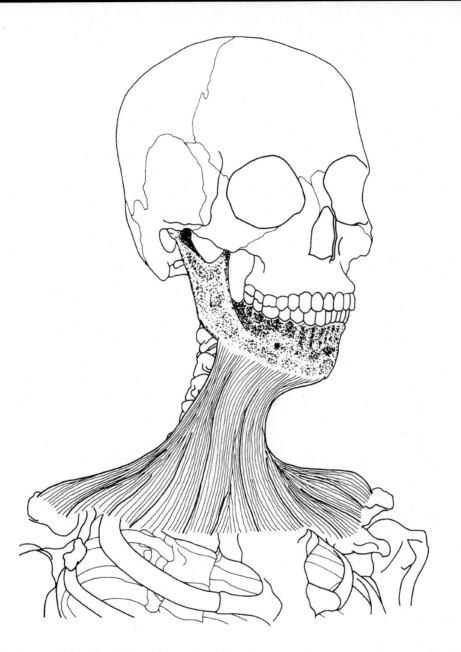

Three-quarter frontal view

■ **Origin**
Subcutaneous fascia of upper one-fourth of chest just below the clavicle

■ **Insertion**
Subcutaneous fascia and muscles of chin and jaw, mandible

■ **Action**
Depresses and draws lower lip laterally, draws up skin of chest, depresses mandible

■ **Nerve**
Cervical branch of facial nerve

DIGASTRICUS

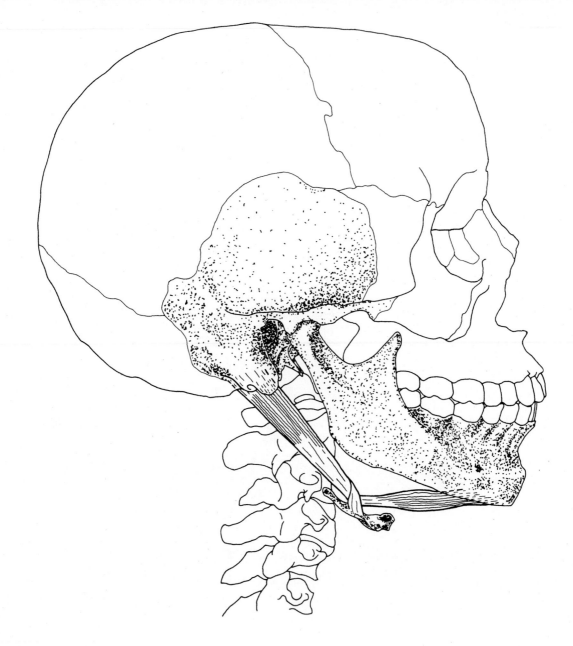

Lateral view

■ Origin

Posterior belly—mastoid notch of temporal bone

Anterior belly—inner side of inferior border of mandible near symphysis

■ Insertion

Intermediate tendon attached to hyoid bone

■ Action

Raises hyoid bone, depresses mandible, moves hyoid forward or backward

■ Nerve

Anterior belly—mandibular division of trigeminal

Posterior belly—facial nerve

STYLOHYOID

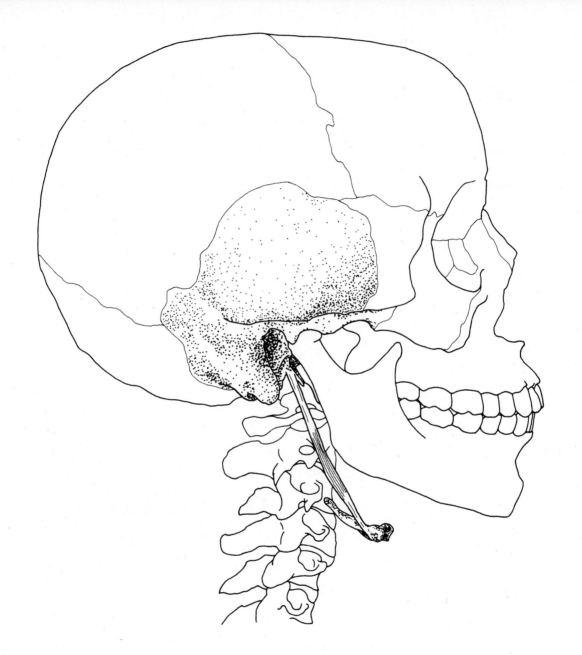

Lateral view

■ **Origin**
Styloid process of temporal bone

■ **Insertion**
Hyoid bone

■ **Action**
Draws hyoid bone backward, elevates tongue

■ **Nerve**
Facial nerve

MYLOHYOID

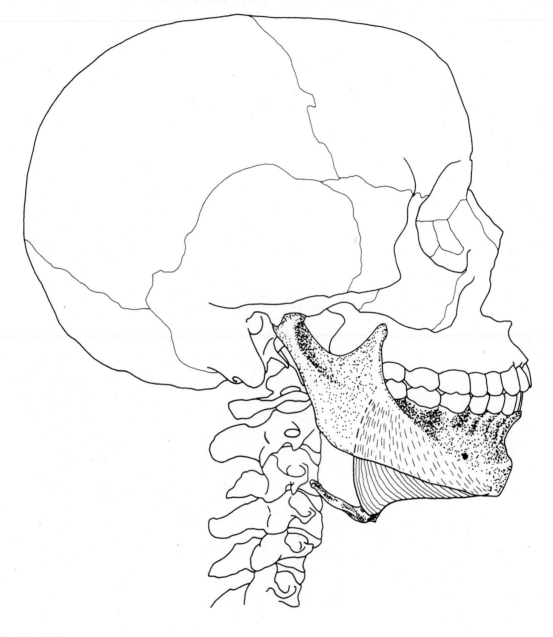

Lateral view

■ **Origin**

Inside surface of mandible from symphysis to molars
(mylohyoid line)

■ **Insertion**

Hyoid bone

■ **Action**

Elevates hyoid bone, raises floor of mouth and tongue

■ **Nerve**

Mandibular division of trigeminal nerve

GENIOHYOID

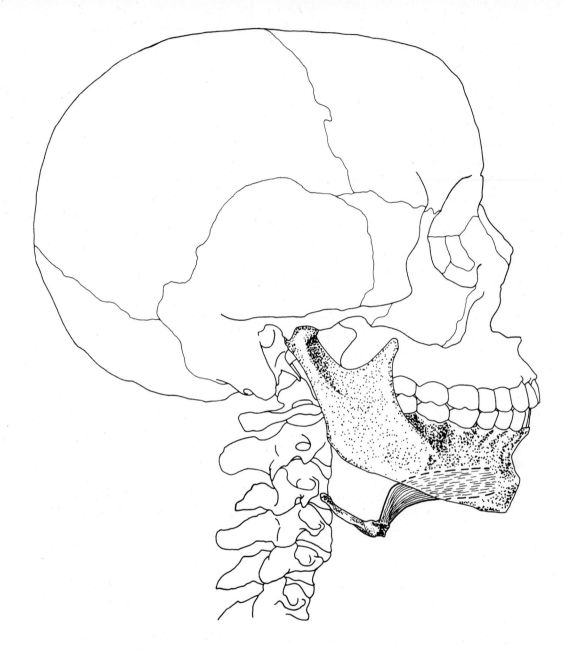

Lateral view

■ Origin
Inferior mental spine on interior medial surface of mandible

■ Insertion
Body of hyoid bone

■ Action
Protrudes hyoid bone and tongue

■ Nerve
Branch of C1 through hypoglossal nerve

STERNOHYOID

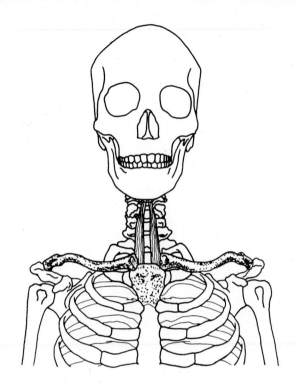

Frontal view

■ **Origin**
Medial end of clavicle, manubrium of sternum

■ **Insertion**
Body of hyoid bone

■ **Action**
Depresses hyoid bone

■ **Nerve**
Ansa cervicalis (C1–C3)

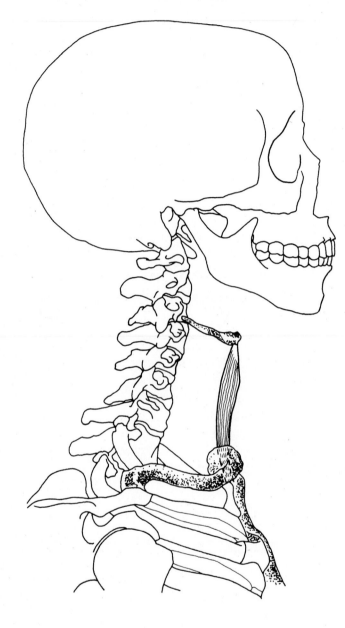

Lateral view

STERNOTHYROID

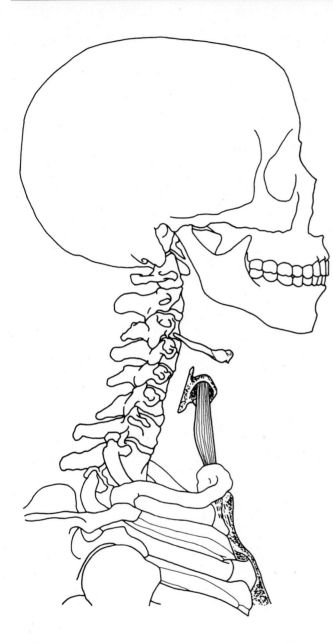

Lateral view

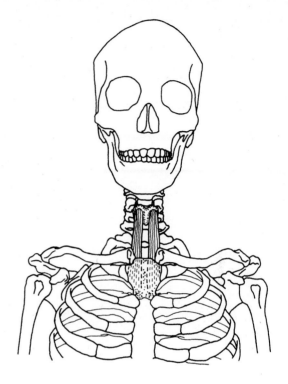

Frontal view

■ **Origin**
Dorsal surface of manubrium of sternum

■ **Insertion**
Lamina of thyroid cartilage

■ **Action**
Depresses thyroid cartilage

■ **Nerve**
Ansa cervicalis (C1–C3)

THYROHYOID

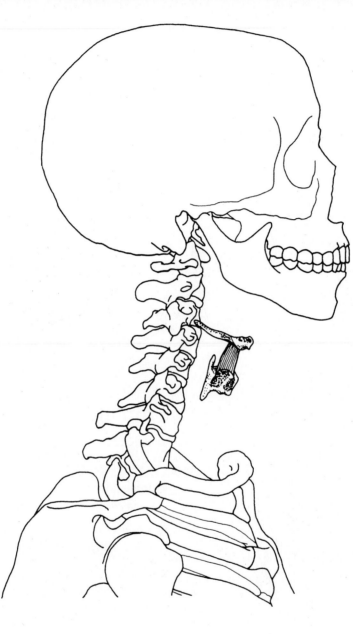

Lateral view

■ Origin
Lamina of thyroid cartilage

■ Insertion
Inferior border of body and greater horn of hyoid bone

■ Action
Depresses hyoid or raises thyroid

■ Nerve
C1 through hypoglossal nerve

OMOHYOID

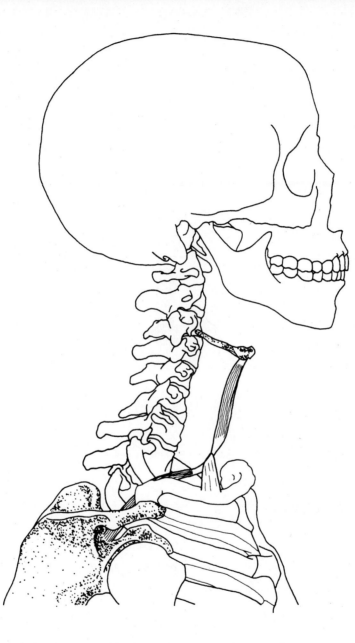

Lateral view

■ Origin
Superior border of scapula

■ Insertion
Inferior belly—bound to clavicle by central tendon

Superior belly—continues to body of hyoid bone

■ Action
Depresses hyoid bone

■ Nerve
Ansa cervicalis (C2, C3)

LONGUS COLLI

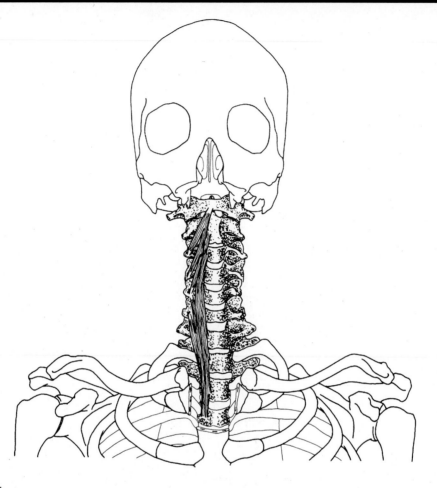

Frontal view
(Mandible and part of maxilla removed)

Superior oblique part

■ Origin
Transverse processes of third, fourth, and fifth cervical vertebrae

■ Insertion
Anterior arch of atlas

Inferior oblique part

■ Origin
Anterior surface of bodies of first two or three thoracic vertebrae

■ Insertion
Transverse processes of fifth and sixth cervical vertebrae

Vertical part

■ Origin
Anterior surfaces of bodies of upper three thoracic and lower three cervical vertebrae

■ Insertion
Anterior surfaces of the second, third, and fourth cervical vertebrae

■ Action
All three parts flex cervical vertebrae

■ Nerve
C2–C7

Note: Cervical hyperextension injuries (whiplash) may strain these muscles and/or sprain the anterior ligaments of vertebrae.

LONGUS CAPITIS

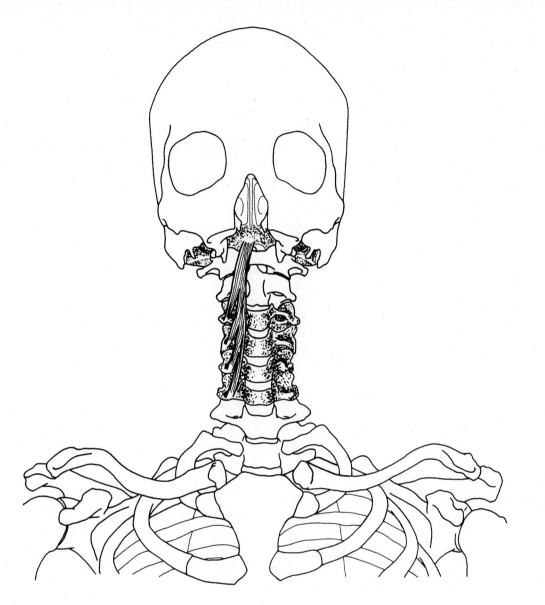

Frontal view
(Mandible and part of maxilla removed)

■ Origin
Anterior tubercles of transverse processes of third through sixth cervical vertebrae

■ Insertion
Occipital bone anterior to foramen magnum

■ Action
Acting together (bilaterally)—flex head

Acting on one side only—rotate head

■ Nerve
C1–C3

RECTUS CAPITIS ANTERIOR

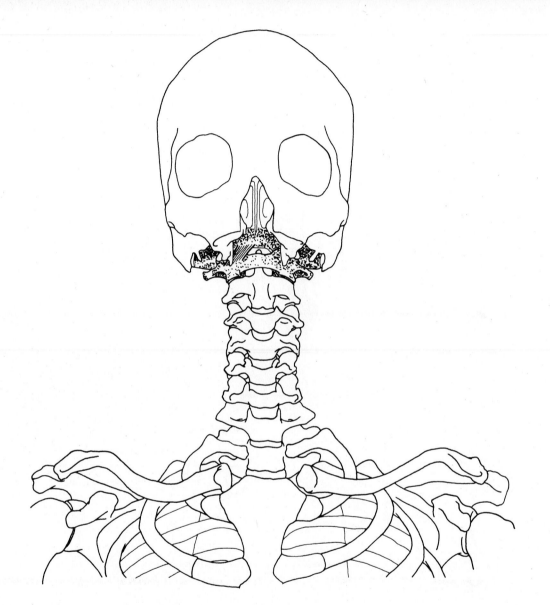

Frontal view
(Mandible and part of maxilla removed)

■ **Origin**
Anterior base of transverse process of atlas

■ **Insertion**
Occipital bone anterior to foramen magnum

■ **Action**
Flexes head

■ **Nerve**
C2, C3

RECTUS CAPITIS LATERALIS

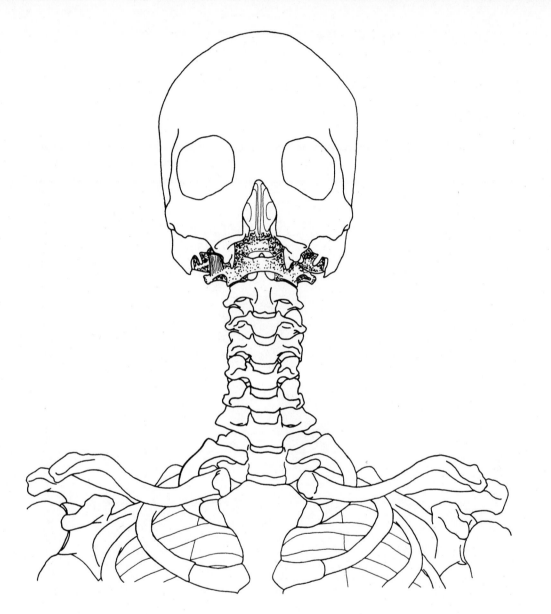

Frontal view
(Mandible and part of maxilla removed)

■ **Origin**
Transverse process of atlas

■ **Insertion**
Jugular process of occipital bone

■ **Action**
Bends head laterally

■ **Nerve**
C2, C3

SCALENUS ANTERIOR

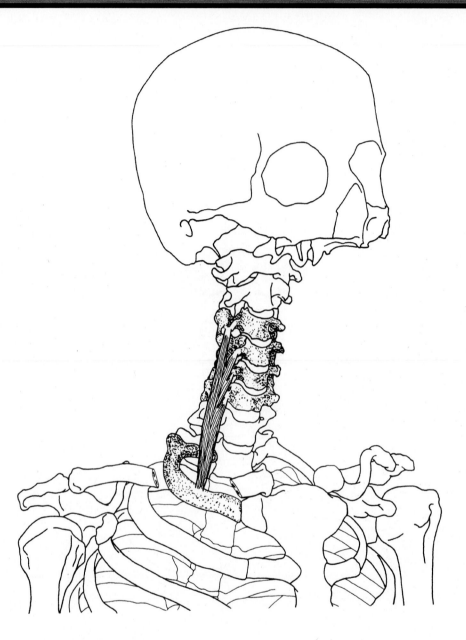

Three-quarter frontal view
(Mandible and part of maxilla removed)

■ Origin
Transverse processes of third through sixth cervical vertebrae

■ Insertion
Inner border of first rib (scalene tubercle)

■ Action
Raises first rib (respiratory inspiration); acting together, they flex neck; acting on one side, they laterally flex, rotate neck

■ Nerve
Ventral rami of cervical nerves

SCALENUS MEDIUS

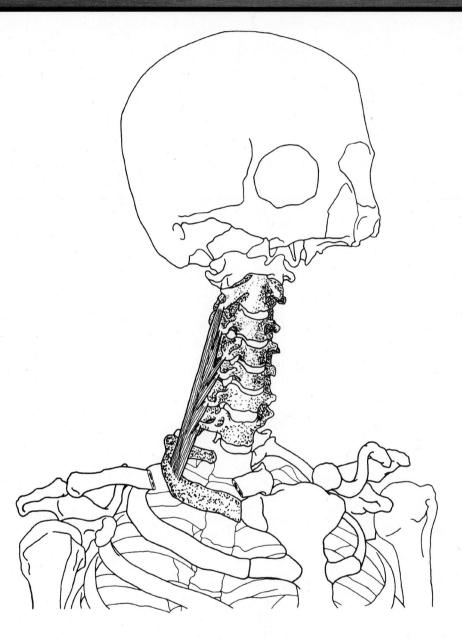

Three-quarter frontal view
(Mandible and part of maxilla removed)

■ **Origin**
Transverse processes of lower six cervical vertebrae
(C2–C7)

■ **Insertion**
Upper surface of first rib

■ **Action**
Raises first rib (respiratory inspiration); acting together,
they flex neck; acting on one side, they laterally flex,
rotate neck

■ **Nerve**
Ventral rami of cervical nerves

SCALENUS POSTERIOR

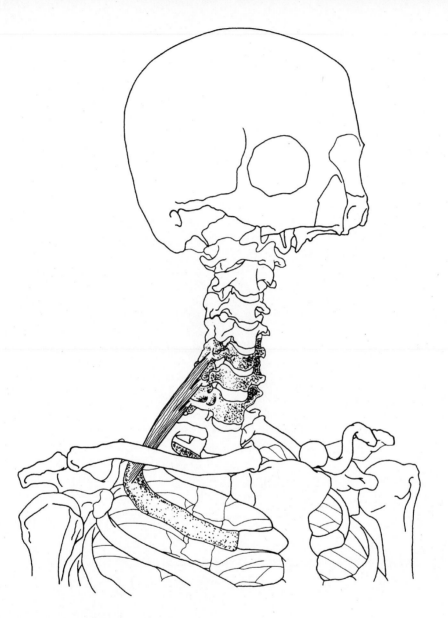

Three-quarter frontal view
(Mandible and part of maxilla removed)

■ **Origin**
Transverse processes of lower two or three cervical vertebrae (C5–C7)

■ **Insertion**
Outer surface of second rib

■ **Action**
Raises second rib (respiratory inspiration); acting together, they flex neck; acting on one side, they laterally flex, rotate neck

■ **Nerve**
Ventral rami of lower cervical nerves

RECTUS CAPITIS POSTERIOR MAJOR

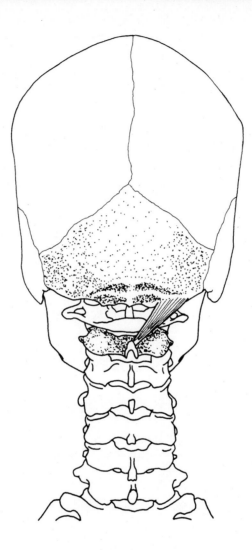

Posterior skull and cervical vertebrae

■ **Origin**
Spinous process of axis

■ **Insertion**
Lateral portion of inferior nuchal line of occipital bone

■ **Action**
Extends and rotates head

■ **Nerve**
Suboccipital nerve

RECTUS CAPITIS POSTERIOR MINOR

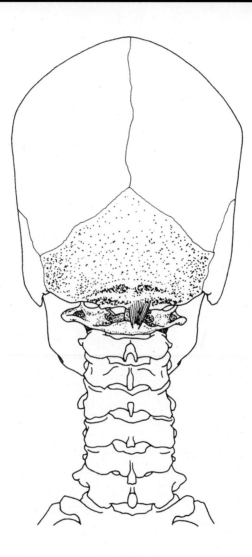

Posterior skull and cervical vertebrae

■ **Origin**
Posterior arch of atlas

■ **Insertion**
Medial portion of inferior nuchal line of occipital bone

■ **Action**
Extends head

■ **Nerve**
Suboccipital nerve

OBLIQUUS CAPITIS INFERIOR

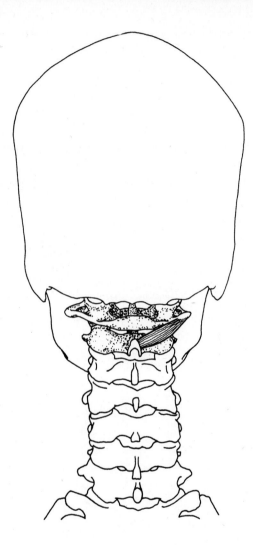

Posterior skull and cervical vertebrae

■ **Origin**
Spinous process of axis

■ **Insertion**
Transverse process of atlas

■ **Action**
Rotates atlas

■ **Nerve**
Suboccipital nerve

OBLIQUUS CAPITIS SUPERIOR

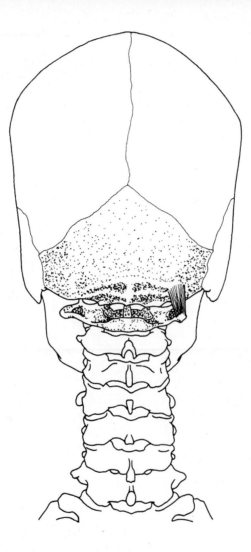

Posterior skull and cervical vertebrae

■ Origin
Transverse process of atlas

■ Insertion
Occipital bone between inferior and superior nuchal lines

■ Action
Extends and bends head laterally

■ Nerve
Suboccipital nerve

CHAPTER FIVE

Muscles of the Trunk

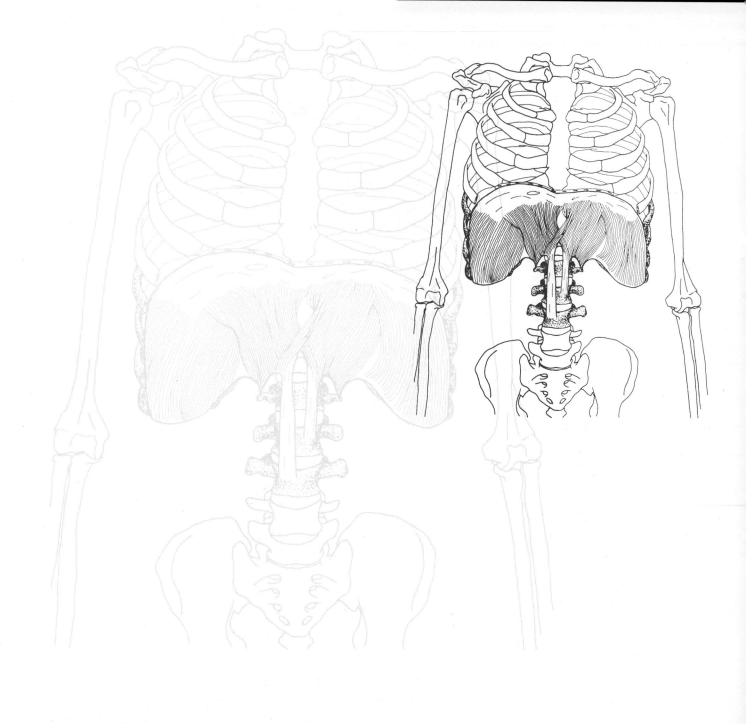

SPLENIUS CAPITIS

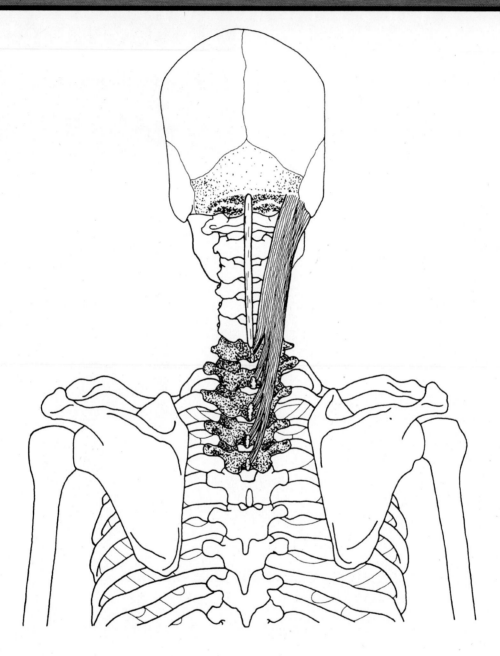

Posterior skull, neck, and back

■ Origin
Lower part of ligamentum nuchae, spinous processes of seventh cervical vertebra (C7) and upper three or four thoracic vertebrae (T1–T4)

■ Insertion
Mastoid process of temporal bone and lateral part of superior nuchal line

■ Action
Acting together, they extend, hyperextend head, neck; acting on one side, they laterally flex, rotate head, neck

■ Nerve
Lateral branches of dorsal primary divisions of middle and lower cervical nerves

SPLENIUS CERVICIS

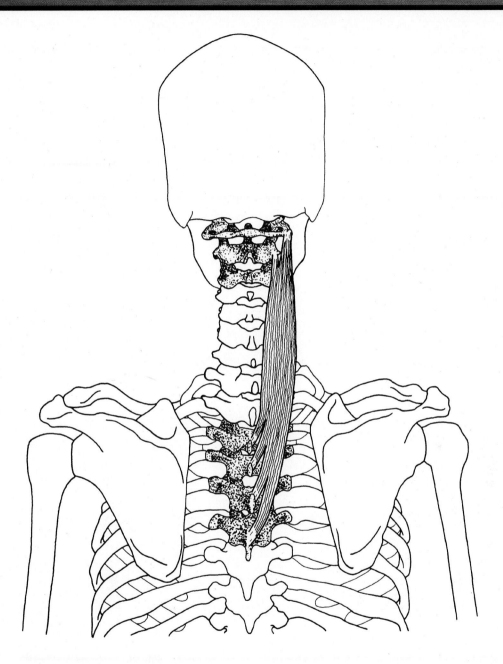

Posterior skull, neck, and back

■ Origin
Spinous processes of third through sixth thoracic vertebrae (T3–T6)

■ Insertion
Transverse processes of upper two or three cervical vertebrae (C1–C3)

■ Action
Acting together, they extend, hyperextend head, neck; acting on one side, they laterally flex, rotate head, neck

■ Nerve
Lateral branches of dorsal primary divisions of middle and lower cervical nerves

ERECTOR SPINAE

Iliocostalis cervicis

■ Origin
Angles of third through sixth ribs

■ Insertion
Transverse processes of fourth, fifth, and sixth cervical vertebrae

■ Action
Extension, lateral flexion of vertebral column

■ Nerve
Dorsal primary divisions of spinal nerves

Iliocostalis thoracis

■ Origin
Angles of lower six ribs medial to iliocostalis lumborum

■ Insertion
Angles of upper six ribs and transverse process of seventh cervical vertebra

■ Action
Extension, lateral flexion of vertebral column, rotates ribs for forceful inspiration

■ Nerve
Dorsal primary divisions of spinal nerves

Iliocostalis lumborum

■ Origin
Medial and lateral sacral crests and medial part of iliac crests

■ Insertion
Angles of lower six ribs

■ Action
Extension, lateral flexion of vertebral column, rotates ribs for forceful inspiration

■ Nerve
Dorsal primary divisions of spinal nerves

*The erector spinae (sacrospinalis) is a complex of three sets of muscles: iliocostalis, longissimus, and spinalis. The origin of this group is the medial and lateral sacral crests, the medial part of iliac crests, and the spinous processes and supraspinal ligament of lumbar and eleventh and twelfth thoracic vertebrae.

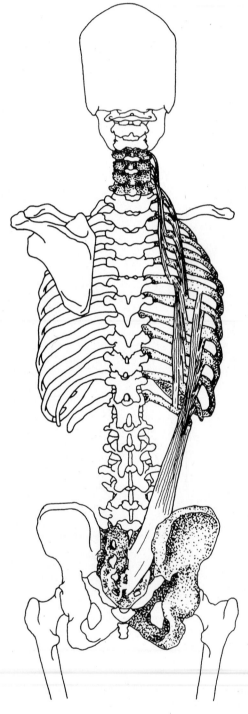

Trunk—dorsal view

ERECTOR SPINAE

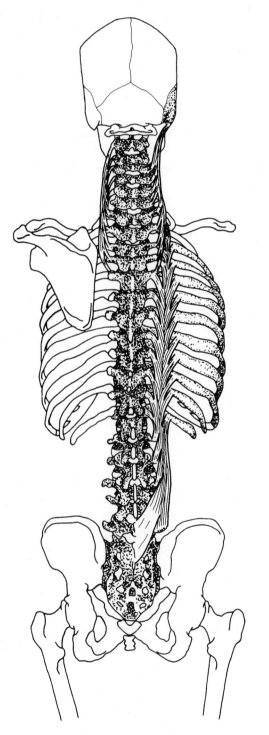

Trunk—dorsal view

Longissimus capitis

■ Origin
Transverse processes of upper five thoracic vertebrae (T1–T5), articular processes of lower three cervical vertebrae (C5–C7)

■ Insertion
Posterior part of mastoid process of temporal bone

■ Action
Extends and rotates head

■ Nerve
Dorsal primary divisions of middle and lower cervical nerves

Longissimus cervicis

■ Origin
Transverse processes of upper four or five thoracic vertebrae (T1–T5)

■ Insertion
Transverse processes of second through sixth cervical vertebrae

■ Action
Extension, lateral flexion of vertebral column

■ Nerve
Dorsal primary divisions of spinal nerves

Longissimus thoracis

■ Origin
Medial and lateral sacral crests, spinous processes and supraspinal ligament of lumbar and eleventh and twelfth thoracic vertebrae, and medial part of iliac crests

■ Insertion
Transverse processes of all thoracic vertebrae, between tubercles and angles of lower nine or ten ribs

■ Action
Extension, lateral flexion of vertebral column, rotates ribs for forceful inspiration

■ Nerve
Dorsal primary divisions of spinal nerves

ERECTOR SPINAE

Spinalis capitis
(Medial part of semispinalis capitis)

Spinalis cervicis

■ Origin
Ligamentum nuchae, spinous process of seventh cervical vertebra

■ Insertion
Spinous process of axis

■ Action
Extends vertebral column

■ Nerve
Dorsal primary divisions of spinal nerves

Spinalis thoracis

■ Origin
Spinous processes of lower two thoracic (T11, T12) and upper two lumbar (L1, L2) vertebrae

■ Insertion
Spinous processes of upper thoracic vertebrae (T1–T8)

■ Action
Extends vertebral column

■ Nerve
Dorsal primary divisions of spinal nerves

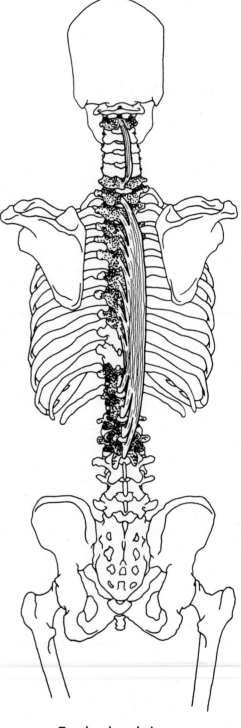

Trunk—dorsal view

TRANSVERSOSPINALIS*

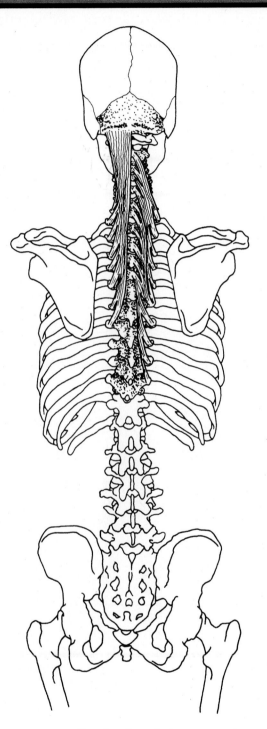

Trunk—dorsal view

Semispinalis capitis
(Medial part is spinalis capitis)

■ Origin
Transverse processes of lower four cervical (C4–C7) and upper six or seven thoracic (T1–T7) vertebrae

■ Insertion
Between superior and inferior nuchal lines of occipital bone

■ Action
Extends and rotates head

■ Nerve
Dorsal primary divisions of spinal nerves

Semispinalis cervicis

■ Origin
Transverse processes of upper five or six thoracic vertebrae (T1–T6)

■ Insertion
Spinous processes of second to fifth cervical vertebrae (C2–C5)

■ Action
Extends and rotates vertebral column

■ Nerve
Dorsal primary divisions of spinal nerves

Semispinalis thoracis

■ Origin
Transverse processes of the sixth through tenth thoracic vertebrae (T6–T10)

■ Insertion
Spinous processes of the lower two cervical (C6, C7) and upper four thoracic (T1–T4) vertebrae

■ Action
Extends and rotates vertebral column

■ Nerve
Dorsal primary divisions of spinal nerves

*The transversospinalis is composed of groups of small muscles generally extending upward from transverse processes to spinous processes of higher vertebrae. They are deep to erector spinae. They include semispinalis, multifidi, and rotatores.

MULTIFIDIS*

Trunk—dorsal view

■ Origin
Sacral region—along sacral foramina up to posterior superior iliac spine

Lumbar region—mammillary processes[†] of vertebrae

Thoracic region—transverse processes

Cervical region—articular processes of lower four vertebrae (C4–C7)

■ Insertion
Spinous process two to four vertebrae superior to origin

■ Action
Extend and rotate vertebral column

■ Nerve
Dorsal primary division of spinal nerves

*Part of transversospinalis.
†Posterior border of superior articular process.

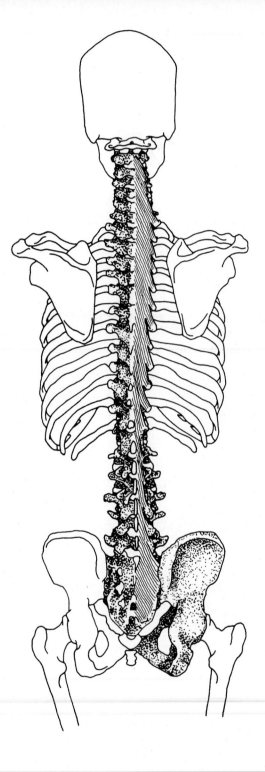

ROTATORES*

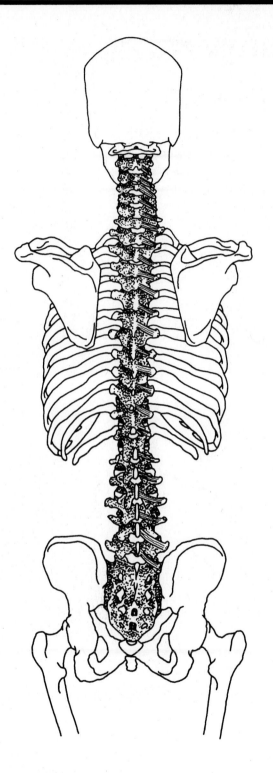

Trunk—dorsal view

■ **Origin**
Transverse process of each vertebra

■ **Insertion**
Base of spinous process of next vertebra above

■ **Action**
Extend and rotate vertebral column

■ **Nerve**
Dorsal primary division of spinal nerves

*Part of transversospinalis.

INTERSPINALES *(Paired on either side of interspinal ligament)*

Trunk—dorsal view

■ Origin
Cervical region—spinous processes of third to seventh cervical vertebrae (C3–C7)

Thoracic region—spinous processes of second to twelfth thoracic vertebrae (T2–T12)

Lumbar region—spinous processes of second to fifth lumbar vertebrae (L2–L5)

■ Insertion
Spinous process of next vertebra superior to origin

■ Action
Extend vertebral column

■ Nerve
Dorsal primary division of spinal nerves

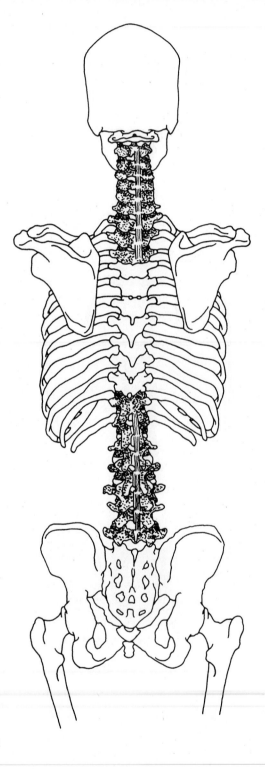

INTERTRANSVERSARII

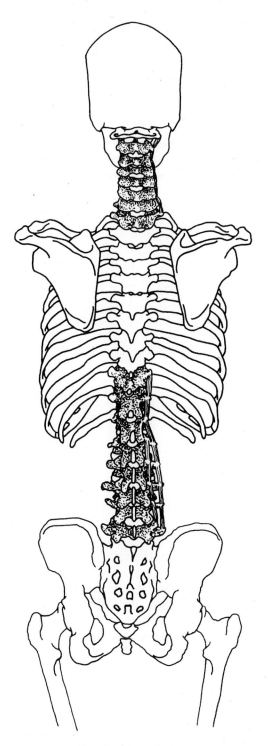

Trunk—dorsal view

Cervical region

Intertransversarii anteriores

- **Origin** Anterior tubercle of transverse processes of vertebrae from first thoracic to axis
- **Insertion** Anterior tubercle of next superior vertebra
- **Action** Lateral flexion of vertebral column
- **Nerve** Ventral primary division of spinal nerves

Intertransversarii posteriores

- **Origin** Posterior tubercle of transverse processes of vertebrae from first thoracic to axis
- **Insertion** Posterior tubercle of next superior vertebra

Thoracic region

- **Origin** Transverse processes of first lumbar to eleventh thoracic vertebrae
- **Insertion** Transverse processes of next superior vertebra

Lumbar region

Intertransversarii laterales

- **Origin** Transverse processes of lumbar vertebrae
- **Insertion** Transverse process of next superior vertebra
- **Action** Lateral flexion of vertebral column
- **Nerve** Ventral primary division of spinal nerves

Intertransversarii mediales

- **Origin** Mammillary process* of each lumbar vertebra
- **Insertion** Accessory process of the next superior lumbar vertebra
- **Action** Lateral flexion of vertebral column
- **Nerve** Dorsal primary division of spinal nerves

*Posterior border of superior articular process.

INTERCOSTALES EXTERNI *(External Intercostal)*

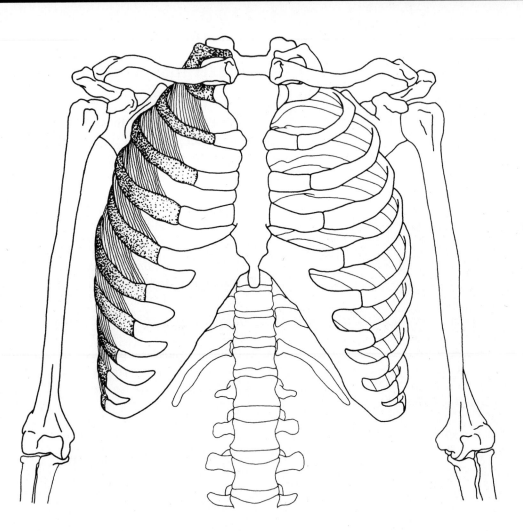

Trunk—anterior view

■ **Origin**
Lower margin of upper eleven ribs

■ **Insertion**
Superior border of rib below (each muscle fiber runs obliquely and inserts toward the costal cartilage)

■ **Action**
Draw ventral part of ribs upward, increasing the volume of the thoracic cavity for inspiration

■ **Nerve**
Intercostal nerves

INTERCOSTALES INTERNI *(Internal Intercostal)*

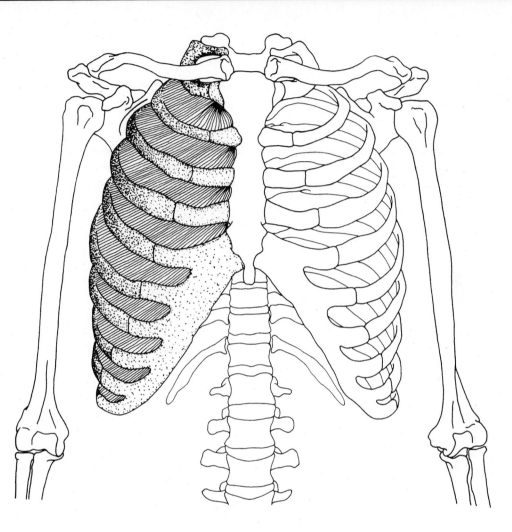

Trunk—anterior view

■ **Origin**
From the cartilages to the angles of the upper eleven ribs

■ **Insertion**
Superior border of the rib below (each muscle fiber runs obliquely and inserts away from the costal cartilage)

■ **Action**
Draw ventral part of ribs downward, decreasing the volume of the thoracic cavity for expiration

■ **Nerve**
Intercostal nerves

SUBCOSTALES

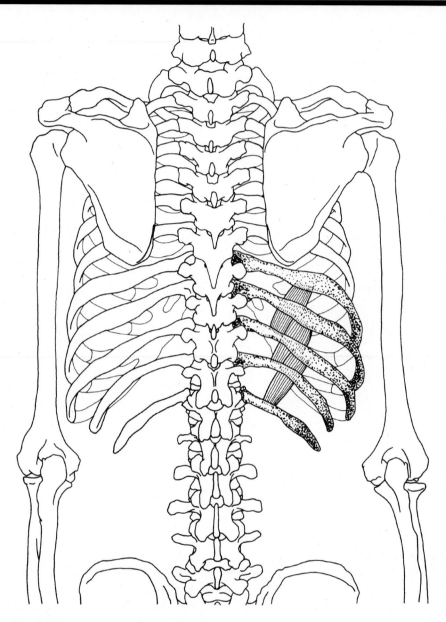

Trunk—dorsal view

■ **Origin**
Inner surface of each rib near its angle

■ **Insertion**
Medially on the inner surface of second or third rib below

■ **Action**
Draw ventral part of ribs downward, decreasing the volume of the thoracic cavity for forceful expiration

■ **Nerve**
Intercostal nerves

Note: These muscles are deep to the internal intercostals. They continue distally between single ribs, where they are known as innermost intercostal muscles.

TRANSVERSUS THORACIS

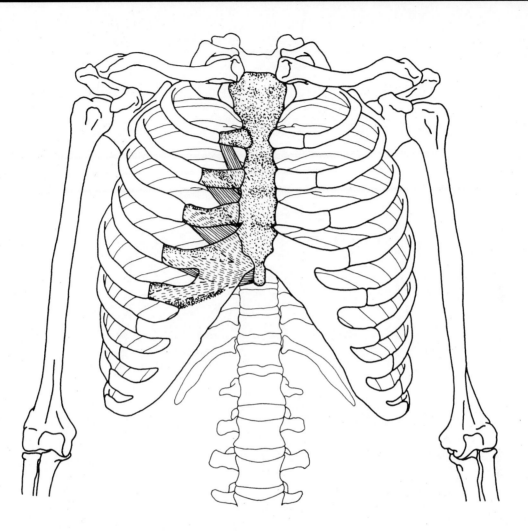

Trunk—anterior view

■ Origin
Inner surface of lower portion of sternum and adjacent costal cartilages

■ Insertion
Inner surfaces of costal cartilages of the second through sixth ribs

■ Action
Draws ventral part of ribs downward, decreasing the volume of the thoracic cavity for forceful expiration

■ Nerve
Intercostal nerves

Note: These muscles are deep to the internal intercostal muscles.

LEVATORES COSTARUM

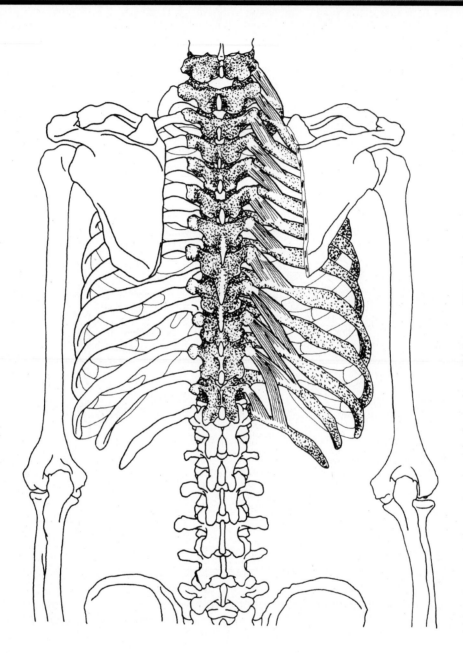

Trunk—dorsal view

■ Origin
Transverse processes of the seventh cervical and the upper eleven thoracic vertebrae

■ Insertion
Laterally to outer surface of next lower rib (lower muscles may cross over one rib)

■ Action
Raises ribs; extends, laterally flexes, and rotates vertebral column

■ Nerve
Intercostal nerves

SERRATUS POSTERIOR SUPERIOR

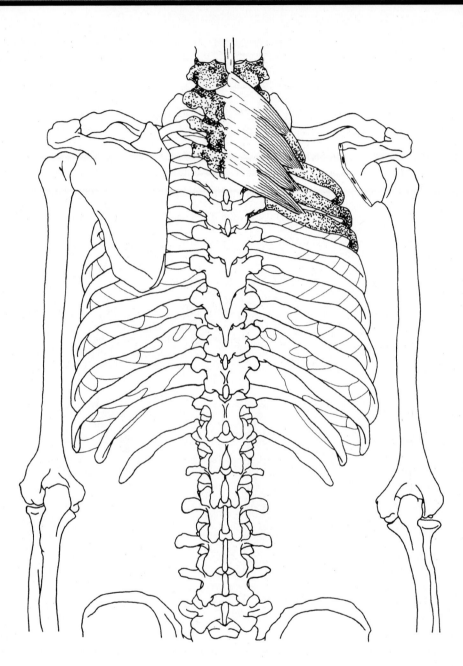

Trunk—dorsal view

■ **Origin**
Ligamentum nuchae, spinous processes of seventh
cervical and first few thoracic vertebrae

■ **Insertion**
Upper borders of the second through fifth ribs lateral to
their angles

■ **Action**
Raises ribs in inspiration

■ **Nerve**
T1–T4

SERRATUS POSTERIOR INFERIOR

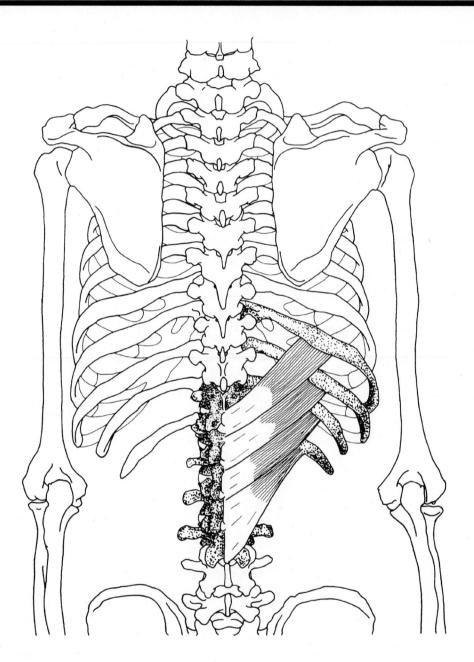

Trunk—dorsal view

■ Origin
Spinous processes of the lower two thoracic and the upper two or three lumbar vertebrae

■ Insertion
Lower borders of bottom four ribs

■ Action
Pulls ribs down, resisting pull of diaphragm

■ Nerve
T9–T12

DIAPHRAGM

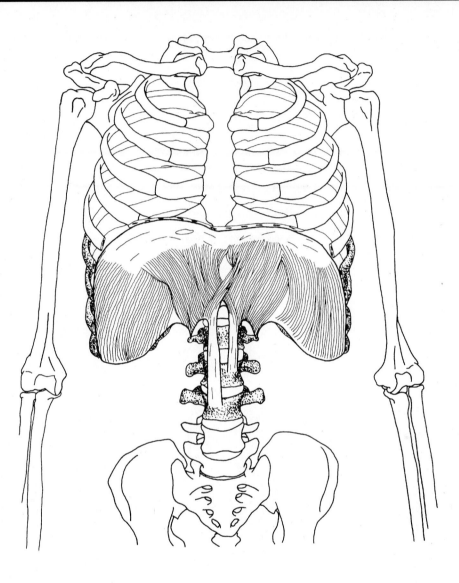

Trunk—anterior view
(Lower costal cartilages removed)

■ Origin
Sternal part—inner part of xiphoid process

Costal part—inner surfaces of lower six ribs and their cartilages

Lumbar part—upper two or three lumbar vertebrae and lateral and medial lumbocostal arches*

■ Insertion
Fibers converge and meet on a central tendon

■ Action
Draws central tendon inferiorly, for inspiration

■ Nerve
Phrenic nerve (C3–C5)

Note: This muscle inserts upon itself. Its action is to change the volume of the thoracic and abdominal cavities.

*These tendinous structures, also known as the medial and lateral arcuate ligaments, allow the diaphragm to bridge the upper parts of the psoas major and quadratus lumborum muscles.

OBLIQUUS EXTERNUS ABDOMINIS *(External Oblique)*

Trunk—lateral view

■ **Origin**
Lower eight ribs

■ **Insertion**
Anterior part of iliac crest, abdominal aponeurosis to linea alba

■ **Action**
Compresses abdominal contents, laterally flexes and rotates vertebral column

■ **Nerve**
Eighth to twelfth intercostal, iliohypogastric, ilioinguinal nerves

■ **Relationships**
Most superficial of the three lateral abdominal muscles

Note: Important in forced expiration, coughing, sneezing.

OBLIQUUS INTERNUS ABDOMINIS *(Internal Oblique)*

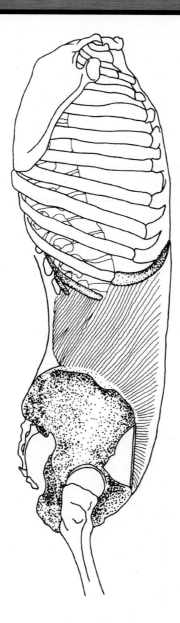

Trunk—lateral view

■ Origin
Lateral half of inguinal ligament, iliac crest, thoracolumbar fascia

■ Insertion
Cartilage of bottom three or four ribs, abdominal aponeurosis to linea alba

■ Action
Compresses abdominal contents, laterally flexes and rotates vertebral column

■ Nerve
Eighth to twelfth intercostal, iliohypogastric, ilioinguinal nerves

■ Relationships
Middle layer of the three lateral abdominal muscles

Note: Important in forced expiration, coughing, sneezing.

CREMASTER

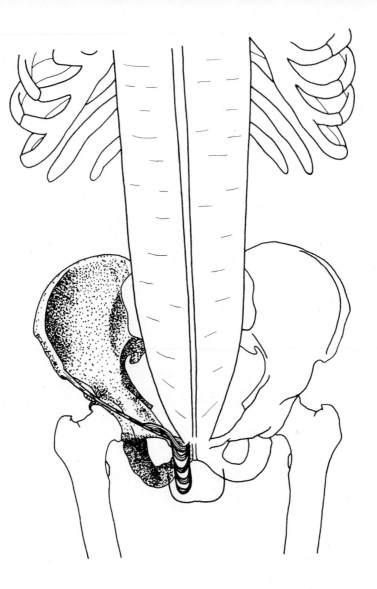

Trunk—anterior view

■ **Origin**
Inguinal ligament

■ **Insertion**
Pubic tubercle, crest of pubis, sheath of rectus abdominis

■ **Action**
Pulls testes toward body

■ **Nerve**
Genital branch of genitofemoral nerve

Note: The cremaster regulates the temperature of the testes, which is important for spermatogenesis.

TRANSVERSUS ABDOMINIS

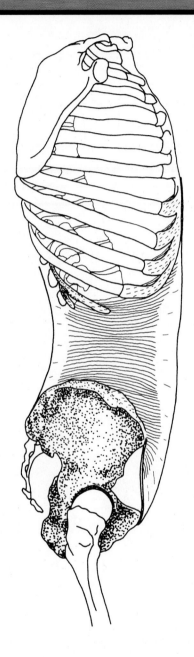

Trunk—lateral view

■ **Origin**
Lateral part of inguinal ligament, iliac crest, thoracolumbar fascia, cartilage of lower six ribs

■ **Insertion**
Abdominal aponeurosis to linea alba

■ **Action**
Compresses abdomen

■ **Nerve**
Seventh to twelfth intercostal, iliohypogastric, ilioinguinal nerves

■ **Relationships**
Deepest of the three lateral abdominal muscles

Note: Important in forced expiration, coughing, sneezing.

RECTUS ABDOMINIS*

Trunk—anterior view

■ Origin
Crest of pubis, pubic symphysis

■ Insertion
Cartilage of fifth, sixth, and seventh ribs, xiphoid process

■ Action
Flexes vertebral column, compresses abdomen

■ Nerve
Seventh through twelfth intercostal nerves

*Tendinous bands divide each rectus into three or four bellies. Each rectus is sheathed in aponeurotic fibers from the lateral abdominal muscles. These fibers meet centrally to form the linea alba.

Note: The pyramidalis is a small, unimportant muscle that extends from the ventral surface of the pubis to the lower part of the linea alba. It is frequently absent.

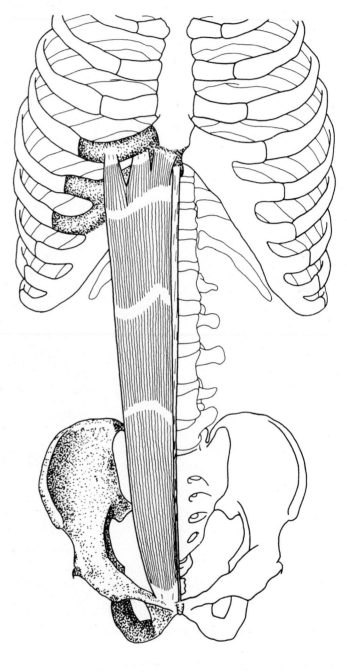

ABDOMINAL MUSCLES

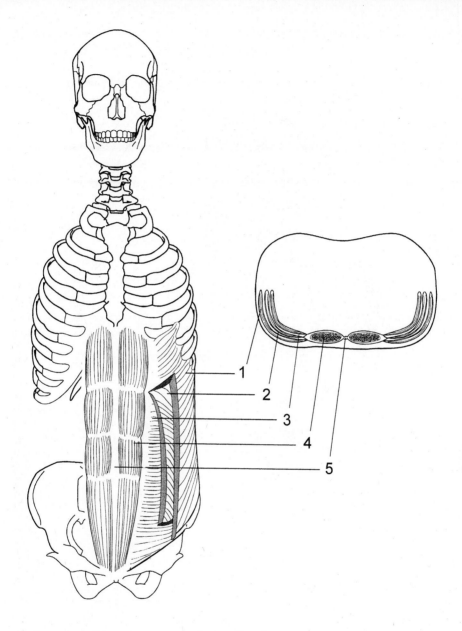

Trunk—anterior and cross-sectional views

1. Obliquus externus abdominis
2. Obliquus internus abdominis
3. Transversus abdominis
4. Rectus abdominis
5. Linea alba

Note: The aponeuroses (broad, flat tendons) of the three lateral abdominal muscles join to form the fascial sheath surrounding the rectus abdominis.

QUADRATUS LUMBORUM

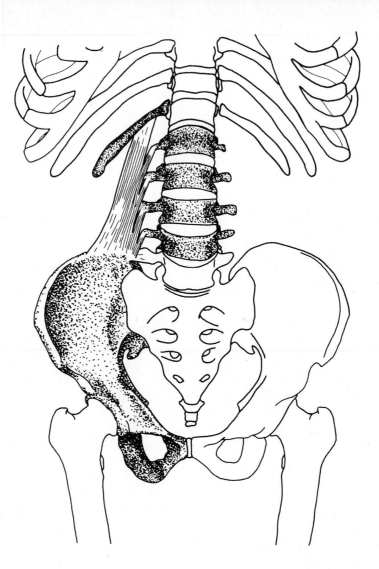

Lower trunk—anterior view

■ Origin
Iliolumbar ligament, iliac crest

■ Insertion
Twelfth rib, transverse processes of upper four lumbar vertebrae

■ Action
Laterally flexes vertebral column, fixes ribs for forced expiration

■ Nerve
T12, L1

Note: Fixation of the ribs may provide a stable attachment of the diaphragm for voice control in singers.

CHAPTER SIX

Muscles of the Shoulder and Arm

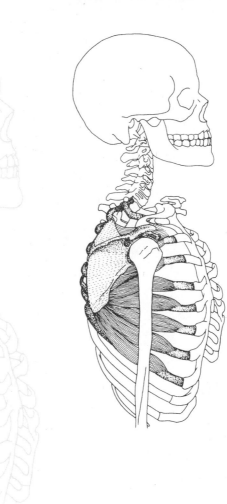

PECTORALIS MAJOR

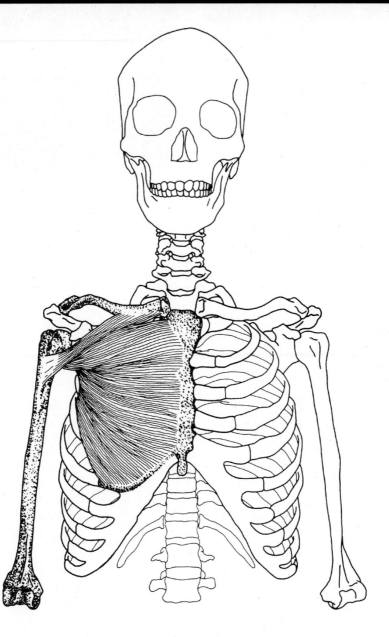

Anterior view

■ Origin
Clavicular part—medial half of the clavicle

Sternocostal part—sternum, upper six costal cartilages, aponeurosis of external oblique

■ Insertion
Lateral lip of intertubercular (bicipital) groove of humerus, crest below greater tubercle of the humerus

■ Action
Both parts adduct, medially rotate arm; clavicular part flexes arm from full extension; sternocostal part extends the flexed arm

■ Nerve
Medial and lateral pectoral nerves (C5–C8, T1)

PECTORALIS MINOR

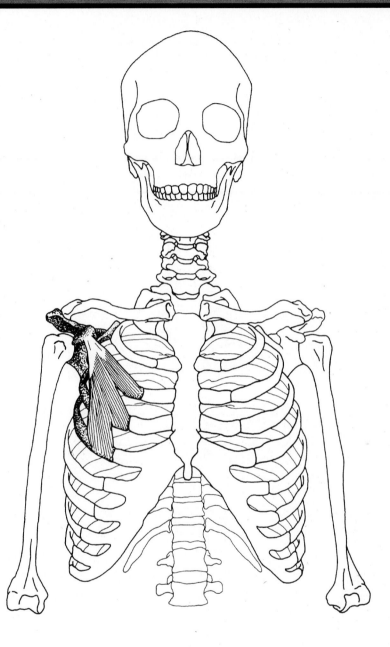

Anterior view

■ **Origin**
External surfaces of the third, fourth, and fifth ribs

■ **Insertion**
Coracoid process of the scapula

■ **Action**
Draws scapula forward and downward, raises ribs* in forced inspiration

■ **Nerve**
Medial pectoral nerve (C8, T1)

■ **Relationships**
Deep to pectoralis major

*Raising the ribs requires stabilization of the scapula by the rhomboids and trapezius.

SUBCLAVIUS

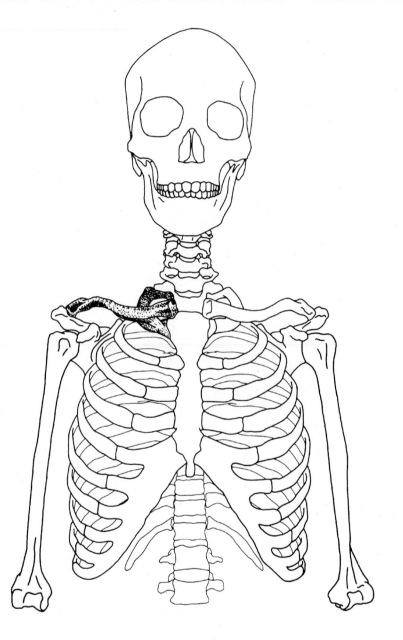

Anterior view

■ **Origin**
Junction of the first rib with its costal cartilage

■ **Insertion**
Groove on the inferior (lower) surface of the clavicle

■ **Action**
Depresses clavicle, draws shoulder forward and downward, steadies clavicle during movements of shoulder girdle

■ **Nerve**
C5, C6

CORACOBRACHIALIS

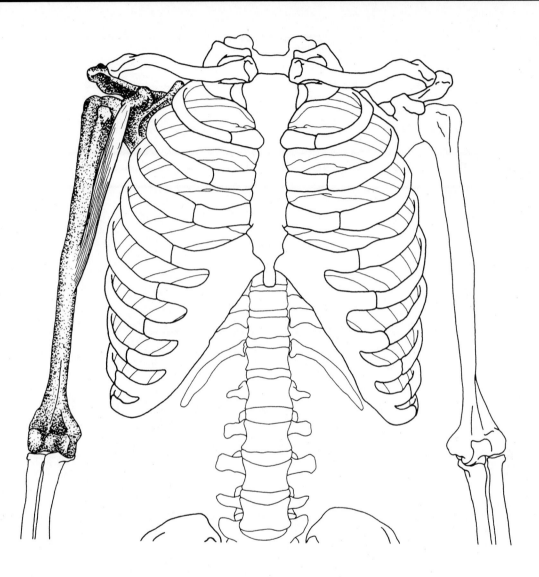

Anterior view

■ Origin
Tip (apex) of the coracoid process of scapula

■ Insertion
Middle third of the medial surface and border of the humerus

■ Action
Weakly adducts arm (flexion unsubstantiated), aids in stabilizing humerus

■ Nerve
Musculocutaneous nerve (C6, C7)

■ Relationships
Deep to short head of biceps

BICEPS BRACHII

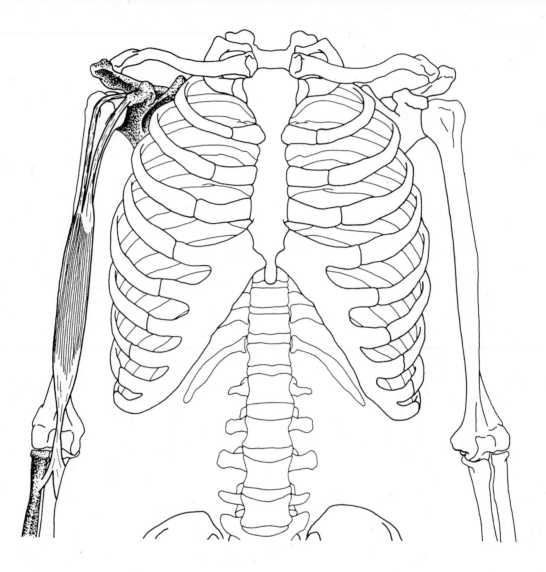

Anterior view

■ Origin

Long head—supraglenoid tubercle of scapula

Short head—coracoid process of scapula

■ Insertion

Tuberosity of radius, bicipital aponeurosis into deep fascia on medial part of forearm

■ Action

Supinates forearm, flexes forearm, weakly flexes arm at shoulder

■ Nerve

Musculocutaneous nerve (C5, C6)

■ Relationships

Long head passes through intertubercular (bicipital) groove, then inside glenohumeral joint capsule

BRACHIALIS

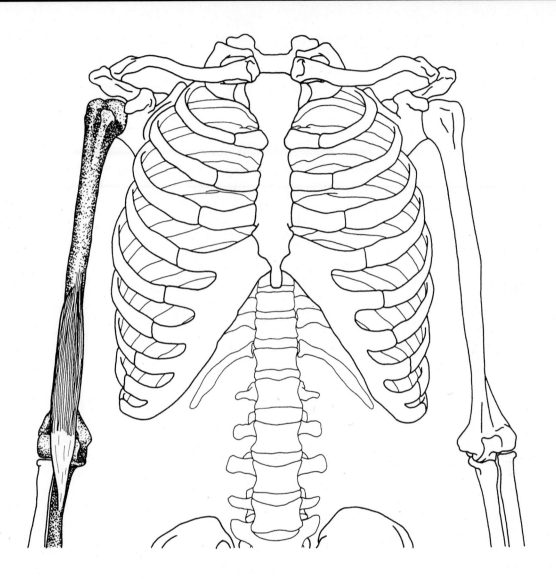

Anterior view

■ **Origin**
Anterior of lower half of humerus

■ **Insertion**
Coronoid process of ulna, tuberosity of ulna

■ **Action**
Flexes forearm

■ **Nerve**
Musculocutaneous nerve (C5, C6)

■ **Relationships**
Deep to biceps brachii

MUSCLES OF THE ANTERIOR CHEST AND ARM

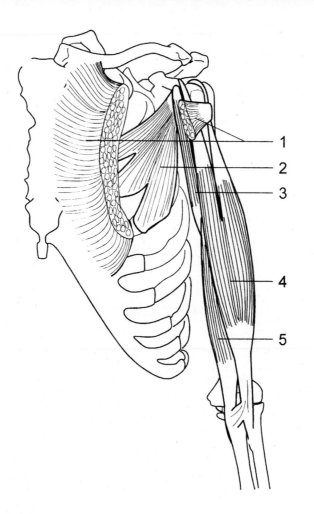

Shoulder—anterior view

1. Pectoralis major (cut)
2. Pectoralis minor
3. Coracobrachialis
4. Biceps brachii
5. Brachialis

TRAPEZIUS

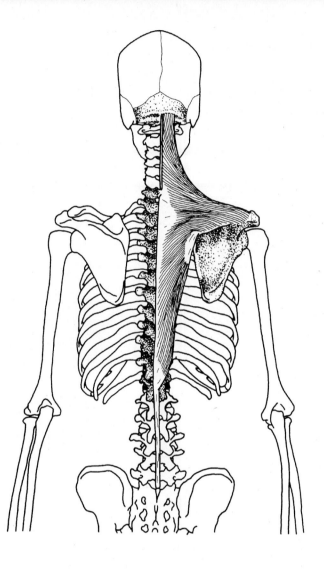

Posterior view

■ Origin
Medial third of superior nuchal line, external occipital protuberance, ligamentum nuchae, spinous processes and supraspinous ligaments of seventh cervical and all thoracic vertebrae

■ Insertion
Upper part—lateral third of clavicle

Middle part—acromion and crest of spine of scapula

Lower part—medial portion of crest of spine of scapula (tubercle)

■ Action
Upper part elevates scapula,* middle part retracts (adducts) scapula, lower part depresses scapula, upper and lower parts together rotate scapula (important in elevating arm)

■ Nerve
Accessory (eleventh cranial), C3, C4

■ Relationships
Most superficial muscle of back

*Upper part stabilizes scapula against downward rotation, as when weight is carried in the hand.

LATISSIMUS DORSI

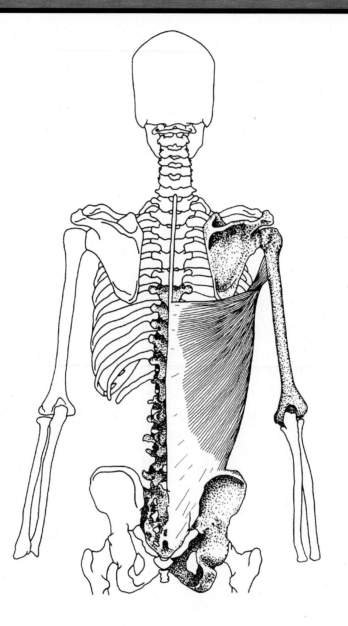

Posterior view

■ Origin

Spinous processes of the lower six thoracic vertebrae, lumbar vertebrae, sacral vertebrae, supraspinal ligament, and posterior part of the iliac crest through the lumbar (thoracolumbar) fascia, lower three or four ribs, inferior angle of the scapula

■ Insertion

Bottom of intertubercular (bicipital) groove

■ Action

Extends, adducts, and medially rotates the arm, draws the shoulder downward and backward, keeps inferior angle of scapula against the chest wall, accessory muscle of respiration

■ Nerve

Thoracodorsal nerve (C6–C8)

Note: This muscle is used for the crawl stroke in swimming.

LEVATOR SCAPULAE

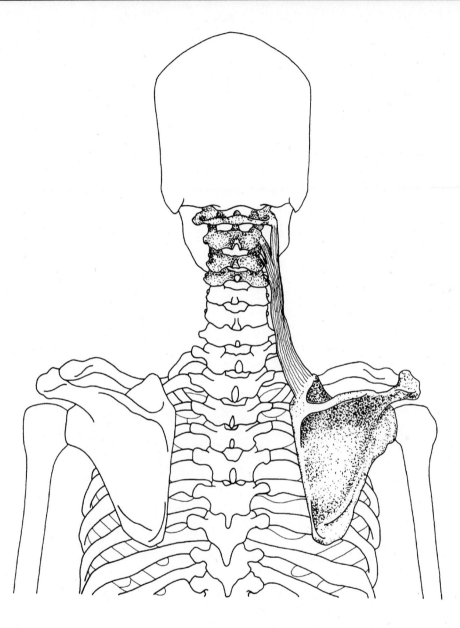

Posterior view

■ Origin
Posterior tubercles of the transverse processes of the first four cervical vertebrae

■ Insertion
Vertebral (medial) border of the scapula at and above the spine

■ Action
Elevates medial border of scapula, rotates scapula to lower the lateral angle, acts with trapezius and rhomboids to pull scapula medially and upward, bends neck laterally

■ Nerve
Dorsal scapular nerve (C5)

RHOMBOID MAJOR

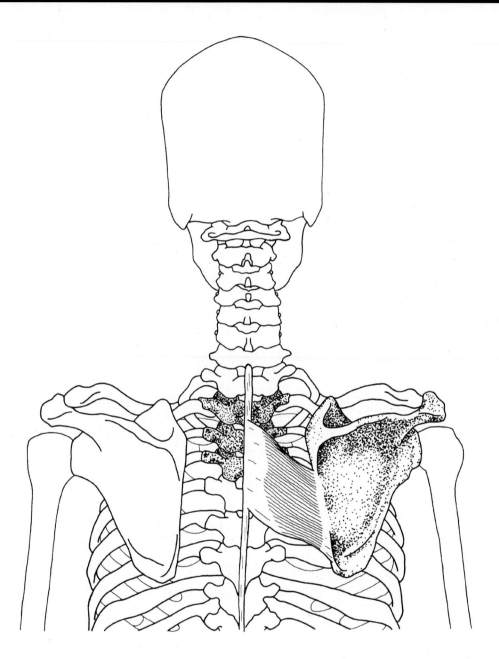

Posterior view

■ Origin
Spines of the second to fifth thoracic vertebrae, supraspinous ligament

■ Insertion
Medial border of the scapula below the spine

■ Action
Retracts and stabilizes scapula, elevates the medial border of the scapula causing downward rotation, assists in adduction of arm

■ Nerve
Dorsal scapular nerve (C5)

RHOMBOID MINOR

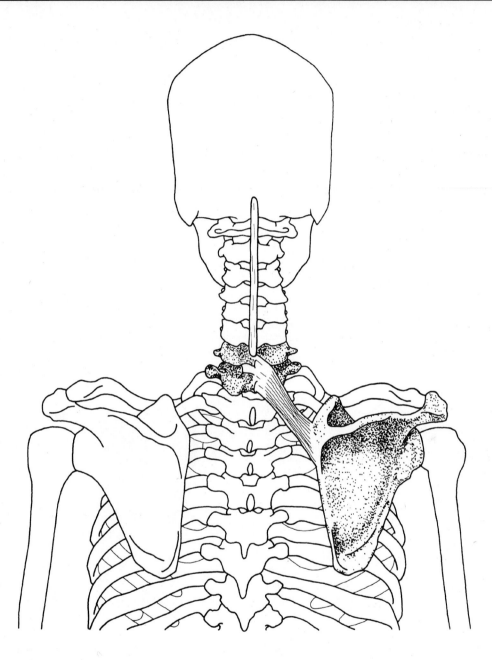

Posterior view

■ Origin
Spines of the seventh cervical and first thoracic vertebrae, lower part of the ligamentum nuchae

■ Insertion
Medial border of the scapula at the root of the spine

■ Action
Retracts and stabilizes scapula, elevates the medial border of the scapula, rotates the scapula to depress the lateral angle (assists in adduction of arm)

■ Nerve
Dorsal scapular nerve (C5)

SERRATUS ANTERIOR

Lateral view

■ Origin
Outer surfaces and superior borders of first eight or nine ribs, and fascia covering first intercostal space

■ Insertion
Anterior surface (costal surface) of the medial border of the scapula

■ Action
Rotates scapula for abduction and flexion of arm, protracts scapula

■ Nerve
Long thoracic nerve (C5–C7)

■ Relationships
Serratus anterior and rhomboids both insert on the medial border of scapula; they are antagonists causing protraction and retraction

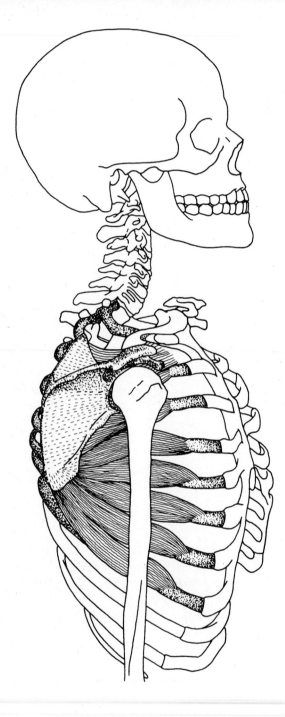

DELTOID

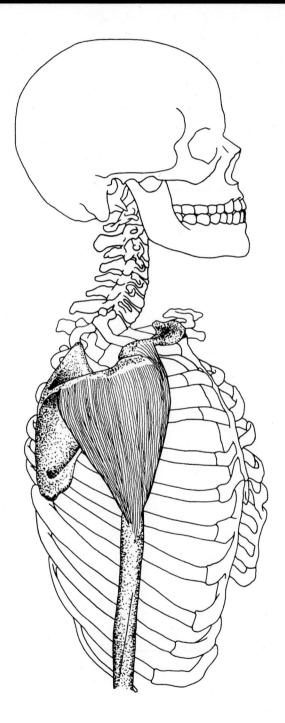

Lateral view

■ Origin
Anterior portion—anterior border and superior surface of the lateral third of the clavicle

Middle portion—lateral border of the acromion process

Posterior portion—lower border of the crest of the spine of the scapula

■ Insertion
Deltoid tuberosity, on the middle of the lateral surface of the shaft of the humerus

■ Action
Anterior portion—flexes and medially rotates arm

Middle portion—abducts arm

Posterior portion—extends and laterally rotates arm

■ Nerve
Axillary nerve (C5, C6)

SUPRASPINATUS *(Rotator Cuff*)*

Lateral view

■ Origin
Supraspinous fossa of scapula

■ Insertion
Upper part of the greater tuberosity of the humerus, capsule of the shoulder joint

■ Action
Aids deltoid in abduction of arm; draws humerus toward glenoid fossa, preventing deltoid from forcing humerus up against acromion; weakly flexes arm

■ Nerve
Suprascapular nerve (C5)

*Supraspinatus, infraspinatus, teres minor, and subscapularis together are called the rotator cuff. They prevent the larger muscles from dislocating the humerus during their actions.

Note: Because of the great range of motion between the humerus and scapula, the joint capsule and its ligaments cannot provide necessary support. The rotator cuff muscles prevent dislocation of the humerus throughout most of the arm's range of motion.

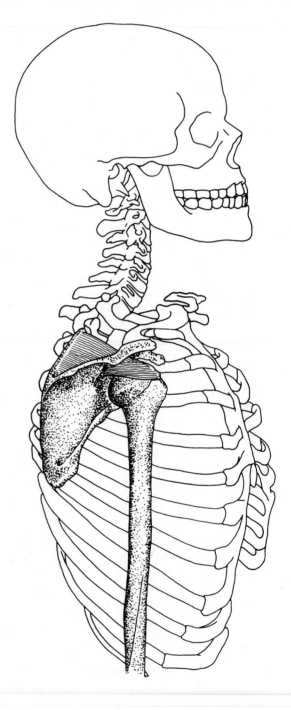

INFRASPINATUS *(Rotator Cuff)*

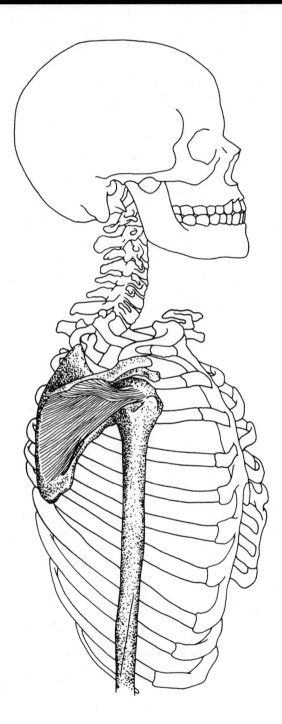

Lateral view

■ Origin
Infraspinous fossa of the scapula

■ Insertion
Middle facet of the greater tuberosity of the humerus, capsule of the shoulder joint

■ Action
Draws humerus toward glenoid fossa, thus resisting posterior dislocation of arm, as in crawling; laterally rotates; abducts arm

■ Nerve
Suprascapular nerve (C5, C6)

TERES MINOR *(Rotator Cuff)*

Lateral view

■ Origin
Upper two-thirds of the dorsal surface of the axillary border of the scapula

■ Insertion
The capsule of the shoulder joint, the lower facet of the greater tuberosity of the humerus

■ Action
Laterally rotates arm, weakly adducts arm, draws humerus toward glenoid fossa

■ Nerve
Axillary nerve (C5)

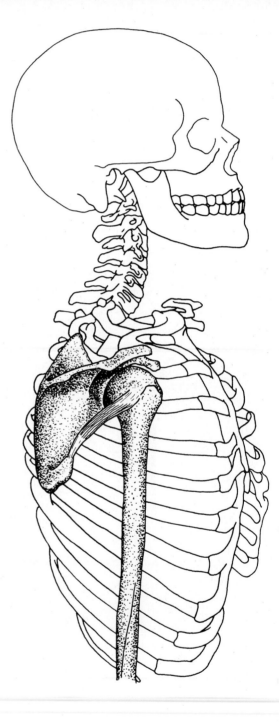

SUBSCAPULARIS *(Rotator Cuff)*

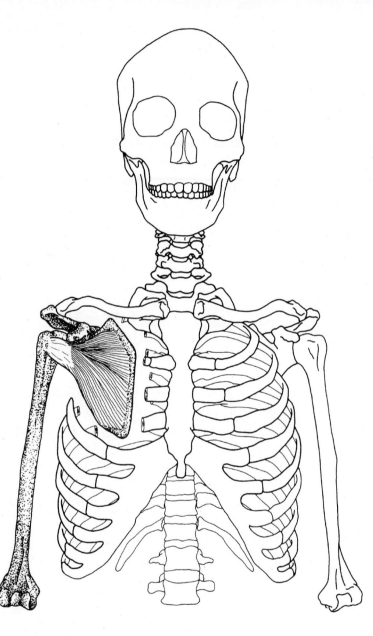

Anterior view
(Upper ribs cut away)

■ **Origin**
Subscapular fossa on the anterior surface of scapula

■ **Insertion**
Lesser tuberosity (tubercle) of the humerus, ventral part of the capsule of the shoulder joint

■ **Action**
Medially rotates arm, stabilizes glenohumeral joint

■ **Nerve**
Upper and lower subscapular nerves (C5, C6)

TERES MAJOR

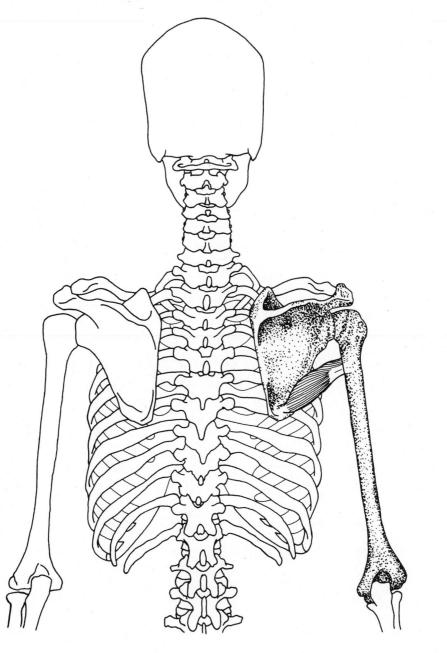

Posterior view

■ Origin
Lower third of the posterior surface of the lateral border of the scapula, near the inferior angle

■ Insertion
Medial lip of the intertubercular (bicipital) groove of the humerus

■ Action
Medially rotates arm, adducts arm, extends arm

■ Nerve
Lower subscapular nerve (C5, C6)

TRICEPS BRACHII

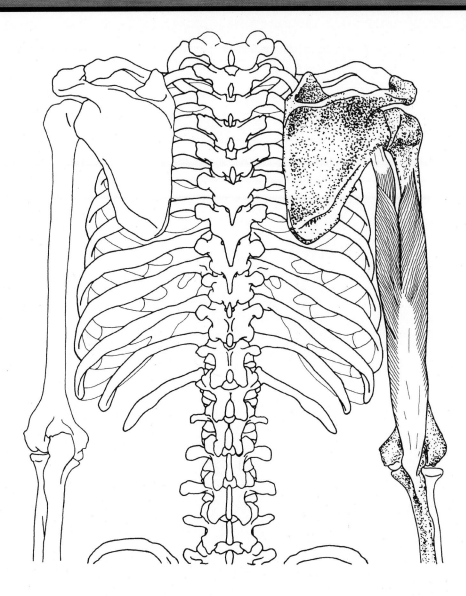

Posterior view

■ Origin

Long head—infraglenoid tubercle of the scapula

Lateral head—upper half of the posterior surface of the shaft of the humerus

Medial head—posterior surface of the lower half of the shaft of the humerus

■ Insertion

Posterior part of olecranon process of the ulna

■ Action

Extends forearm, long head aids in adduction if arm is abducted

■ Nerve

Radial nerve (C7, C8)

Note: The radial nerve comes from the axilla (armpit) and passes along the humerus between the medial and lateral heads. Because it can be compressed against the humerus, it is the most commonly injured peripheral nerve.

ANCONEUS

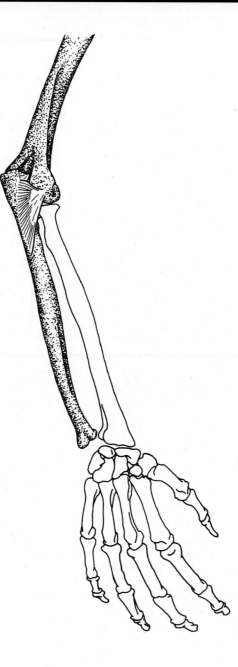

Posterior view of arm

■ Origin
Posterior part of lateral epicondyle of the humerus

■ Insertion
Lateral surface of the olecranon process and posterior surface of ulna

■ Action
Extends forearm (assists triceps)

■ Nerve
Radial nerve (C7, C8)

POSTERIOR BACK, SHOULDER, AND ARM MUSCLES

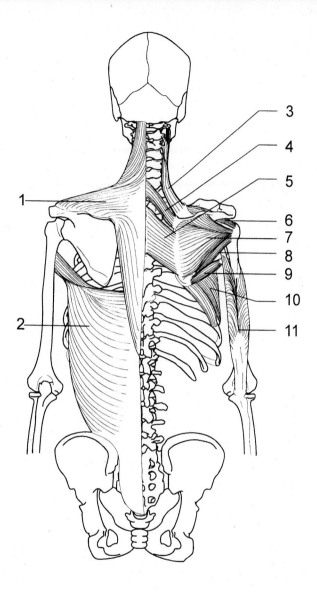

Trunk—posterior view

Superficial layer

1. Trapezius
2. Latissimus dorsi

Deep layer

3. Levator scapulae
4. Rhomboid minor
5. Rhomboid major

6. Supraspinatus (rotator cuff)
7. Infraspinatus (rotator cuff)
8. Teres minor (rotator cuff)
9. Teres major
10. Serratus anterior

Posterior arm

11. Triceps brachii

CHAPTER SEVEN

Muscles of the Forearm and Hand

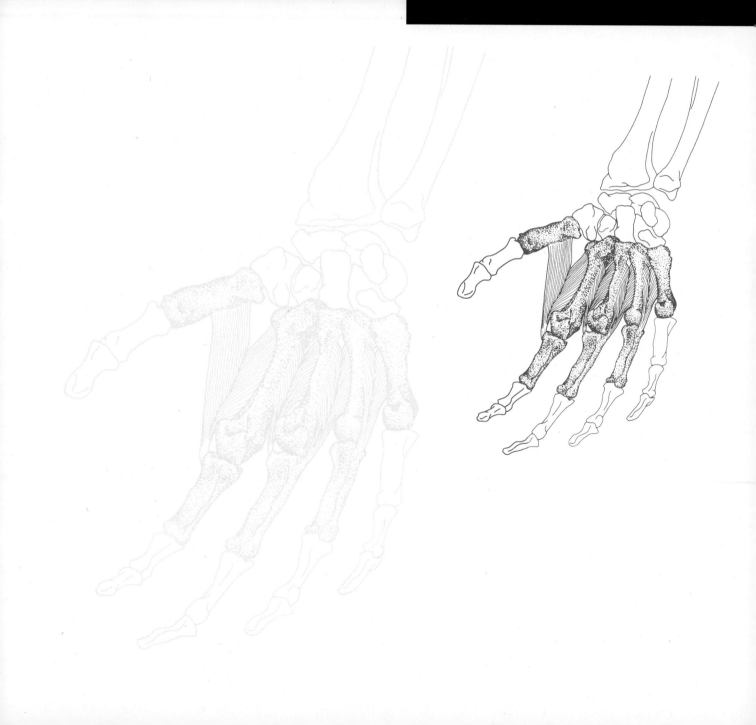

PRONATOR TERES *(Superficial group)*

Forearm—anterior view

■ Origin
Humeral head—medial supracondylar ridge and medial epicondyle of the humerus

Ulnar head—medial border of the coronoid process of the ulna

■ Insertion
Middle of lateral surface of the radius (pronator tuberosity)

■ Action
Pronates and flexes forearm

■ Nerve
Median nerve (C6, C7)

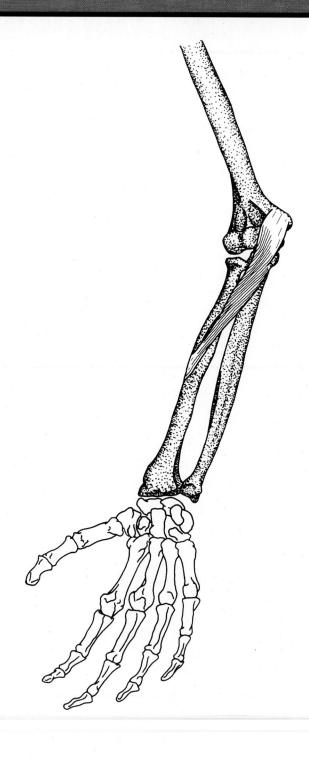

FLEXOR CARPI RADIALIS *(Superficial group)*

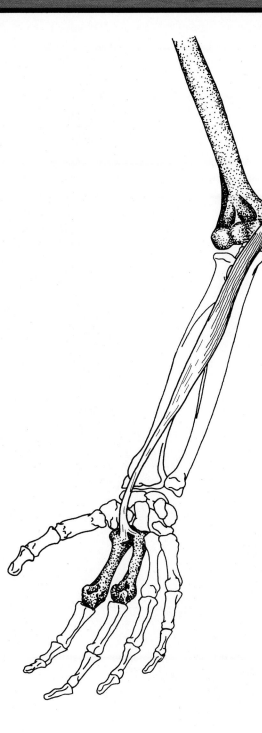

Forearm—anterior view

■ Origin
Medial epicondyle of the humerus through the common tendon

■ Insertion
Front of the bases of the second and third metacarpal bones

■ Action
Flexes hand, synergist in abduction with extensor carpi radialis longus and brevis

■ Nerve
Median nerve (C6, C7)

PALMARIS LONGUS *(Superficial group)*

Forearm—anterior view

■ Origin
Medial epicondyle of the humerus through the common tendon

■ Insertion
Front (central part) of the flexor retinaculum and apex of the palmar aponeurosis

■ Action
Flexes the hand

■ Nerve
Median nerve (C6, C7)

Note: This muscle is absent in about 14% of limbs.

Reference: Agur, Amr: *Grant's Atlas of Anatomy,* 9th ed. Williams & Wilkins, Baltimore, 1991.

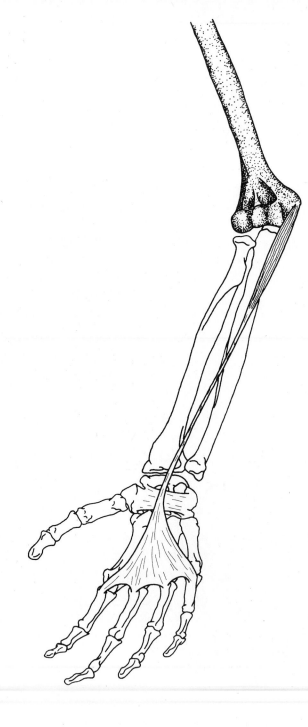

FLEXOR CARPI ULNARIS *(Superficial group)*

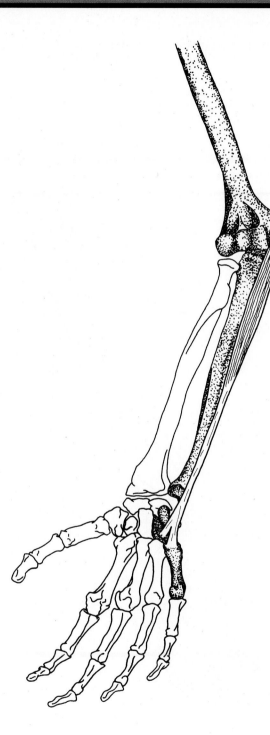

Forearm—anterior view

■ Origin
Humeral head—medial epicondyle of the humerus through the common tendon

Ulnar head—medial margin of olecranon process of ulna, dorsal border of shaft of the ulna

■ Insertion
Pisiform bone, hook of the hamate, and base of the fifth metacarpal bone

■ Action
Flexes hand, synergist in adduction of hand with extensor carpi ulnaris

■ Nerve
Ulnar nerve (C8, T1)

MUSCLES OF THE WRIST

Forearm—anterior view

1. Pronator teres
2. Flexor carpi radialis
3. Palmaris longus
4. Flexor carpi ulnaris
5. Flexor retinaculum
6. Palmar aponeurosis

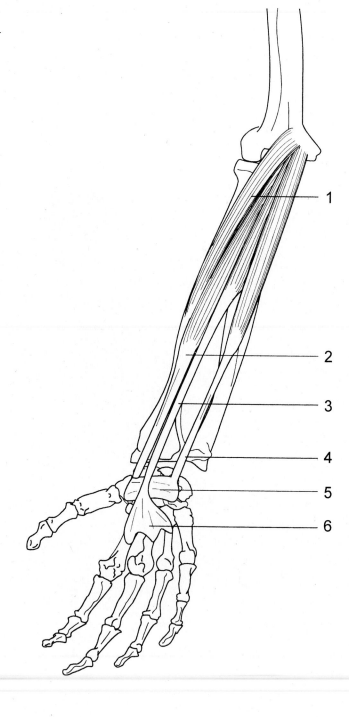

FLEXOR DIGITORUM SUPERFICIALIS

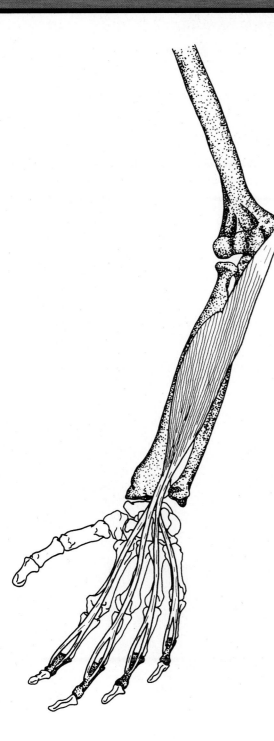

Forearm—anterior view

■ Origin
Humeroulnar head—medial epicondyle of the humerus through common tendon,* medial margin of the coronoid process of ulna

Radial head—anterior surface of shaft of radius

■ Insertion
Four tendons divide into two slips each; slips insert into the sides (margins of the anterior surfaces) of the middle phalanges of four fingers

■ Action
Flexes the middle phalanges of the fingers

■ Nerve
Median nerve (C7, C8, T1)

■ Relationships
Deep to superficial flexors

*See superficial flexors.

Note: The tendons of flexor digitorum superficialis split and attach to the middle phalanx. The tendons of flexor digitorum profundus pass through this split and continue to the distal phalanx.

FLEXOR DIGITORUM PROFUNDUS

Forearm—anterior view

■ Origin
Upper three-fourths of anterior and medial surfaces of shaft of ulna and medial side of the coronoid process, interosseous membrane

■ Insertion
Front of base of distal phalanges of fingers

■ Action
Flexes distal phalanges

■ Nerve
Ulnar nerve supplies the medial half of the muscle (going to the little and ring fingers) (C8, T1)

Anterior interosseous branch of median nerve supplies lateral half (going to index and middle fingers) (C8, T1)

■ Relationships
Deep to flexor digitorum superficialis

Note: Both flexor digitorum muscles and the median nerve pass under the flexor retinaculum (p. 126) in the wrist. When irritated, the synovial sheaths of these muscles can compress the median nerve, causing the sensory and motor deficits known as carpal tunnel syndrome.

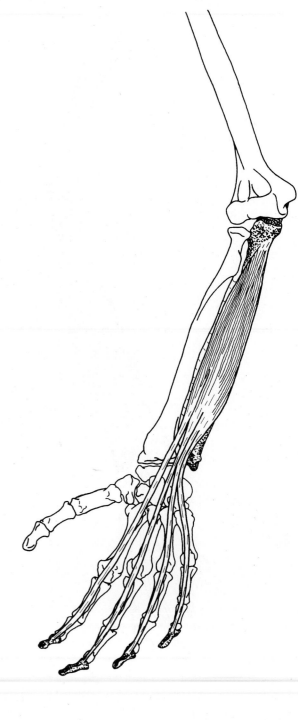

FLEXOR POLLICIS LONGUS

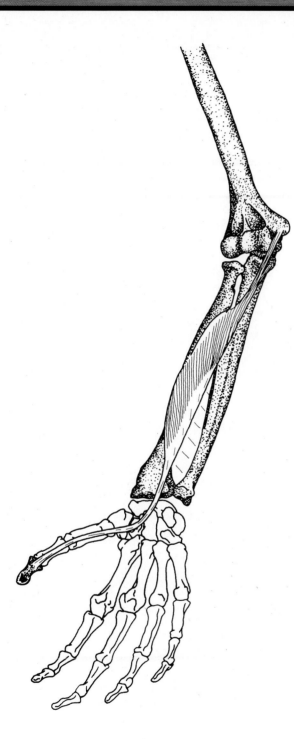

Forearm—anterior view

■ Origin
Middle of anterior surface of shaft of radius, interosseous membrane, medial epicondyle of humerus, and often coronoid process of ulna

■ Insertion
Palmar aspect of base of the distal phalanx of thumb

■ Action
Flexes the thumb

■ Nerve
Anterior interosseous branch of median nerve (C8, T1)

FLEXORS OF THE FINGERS

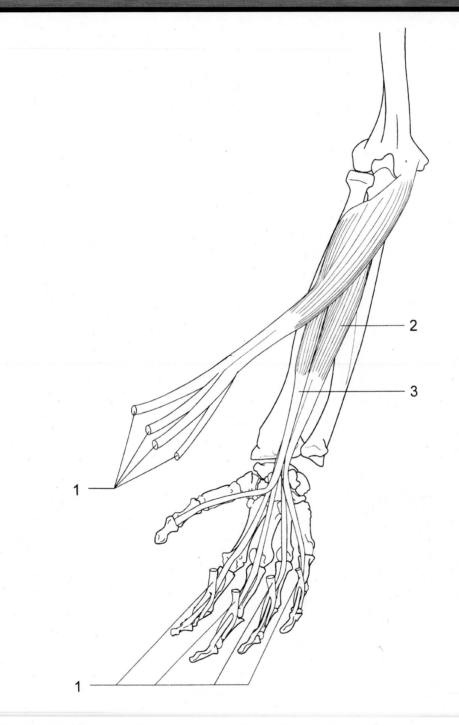

Forearm—anterior view

1. Flexor digitorum superficialis (cut)
2. Flexor digitorum profundus
3. Flexor pollicis longus

PRONATOR QUADRATUS

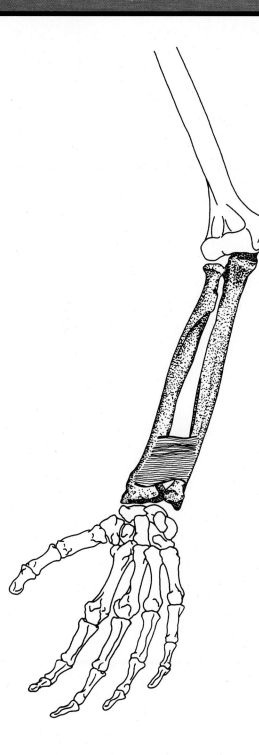

Forearm—anterior view

■ Origin
Anterior surface of distal part of shaft of ulna

■ Insertion
Lower portion of anterior surface of shaft of radius, distal part of lateral border of radius

■ Action
Pronates forearm and hand

■ Nerve
Anterior interosseous branch of median nerve (C8, T1)

■ Relationships
Deepest forearm muscle

BRACHIORADIALIS

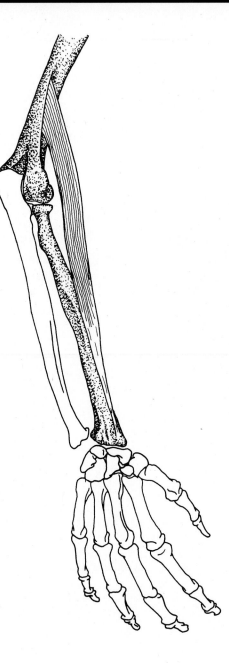

Forearm—dorsal view

■ Origin
Upper two-thirds of lateral supracondylar ridge of humerus

■ Insertion
Base of styloid process and lateral surface of radius

■ Action
Flexes forearm

■ Nerve
Radial nerve (C5, C6)

EXTENSOR CARPI RADIALIS LONGUS

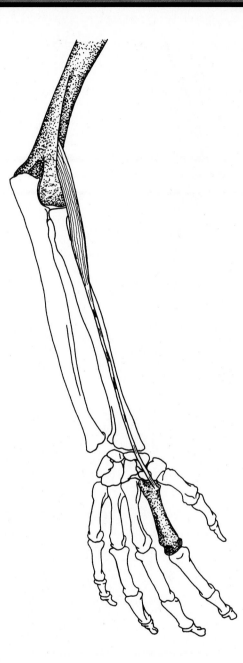

Forearm—dorsal view

■ Origin
Lower third of lateral supracondylar ridge of humerus

■ Insertion
Dorsal surface of the base of the second metacarpal bone

■ Action
Extends hand, synergist in abduction of hand with flexor carpi radialis

■ Nerve
Radial nerve (C6, C7)

EXTENSOR CARPI RADIALIS BREVIS

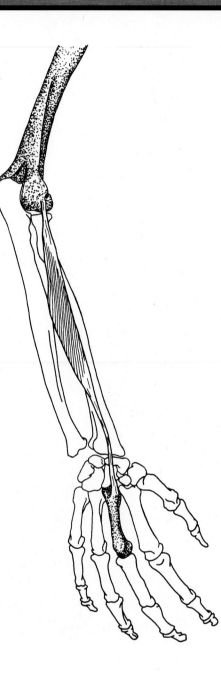

Forearm—dorsal view

■ Origin
Lateral epicondyle of humerus

■ Insertion
Dorsal surface of third metacarpal bone

■ Action
Extends hand, synergist in abduction of hand with flexor carpi radialis

■ Nerve
Radial nerve (C6, C7)

EXTENSOR DIGITORUM COMMUNIS

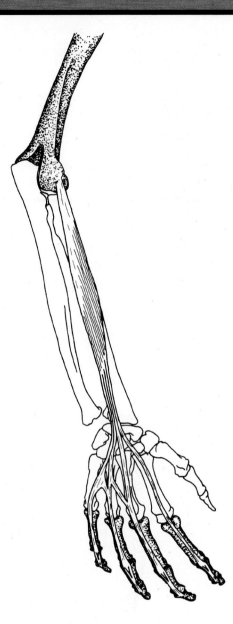

Forearm and hand—dorsal view

■ Origin
Common tendon attached to lateral epicondyle of humerus

■ Insertion
Lateral and dorsal surfaces of all the phalanges of the four fingers

■ Action
Extends the fingers and wrist

■ Nerve
Deep branch of radial nerve (C6–C8)

■ Relationships
Tends to hyperextend the metacarpophalangeal joint causing "claw hand"; its action is balanced by the lumbricales and interossei

EXTENSOR DIGITI MINIMI

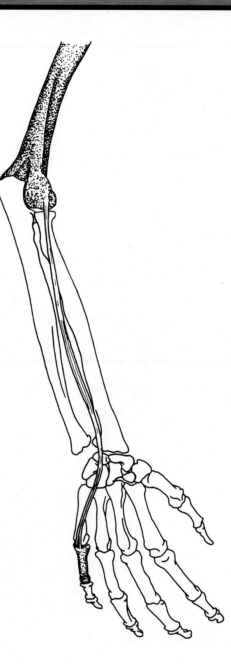

Forearm and hand—dorsal view

■ Origin
Common tendon attached to lateral epicondyle of humerus

■ Insertion
Dorsal surface of base of first phalanx of fifth finger

■ Action
Extends fifth finger

■ Nerve
Radial nerve (C6–C8)

EXTENSOR CARPI ULNARIS

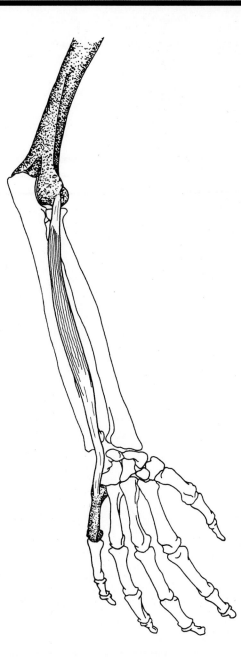

Forearm and hand—dorsal view

■ **Origin**
Common tendon attached to lateral epicondyle of humerus

■ **Insertion**
Dorsal surface of base of fifth metacarpal bone

■ **Action**
Extends hand, synergist in adduction of hand with flexor carpi ulnaris

■ **Nerve**
Radial nerve (C6–C8)

SUPINATOR

Forearm and hand—anterior view

■ **Origin**
Lateral epicondyle of humerus, lateral ligament (radial collateral) of elbow, annular ligament of superior radioulnar joint, supinator crest of ulna

■ **Insertion**
Dorsal and lateral surfaces of upper third of radius

■ **Action**
Supinates forearm

■ **Nerve**
Radial nerve (C6)

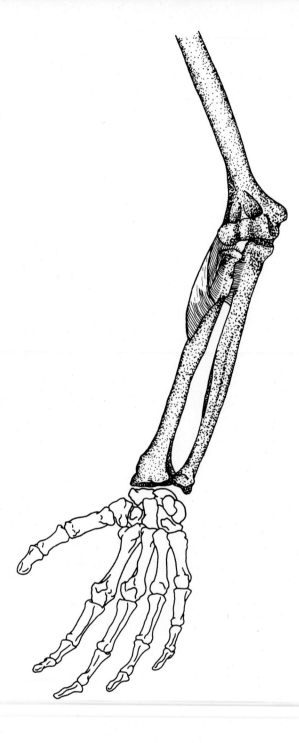

ABDUCTOR POLLICIS LONGUS

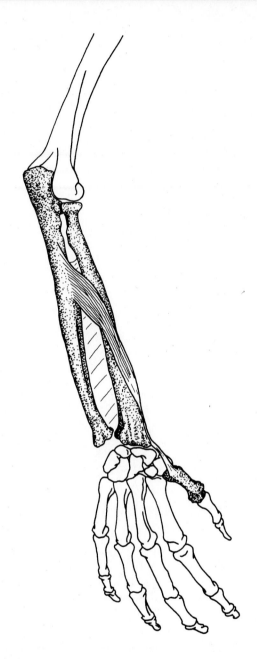

Forearm and hand—dorsal view

■ **Origin**
Posterior (dorsal) surface of shaft of radius, ulna, interosseous membrane

■ **Insertion**
Dorsal surface of base of first metacarpal bone

■ **Action**
Abducts, laterally rotates, and extends thumb; abducts wrist

■ **Nerve**
Radial nerve (C6, C7)

EXTENSOR POLLICIS BREVIS

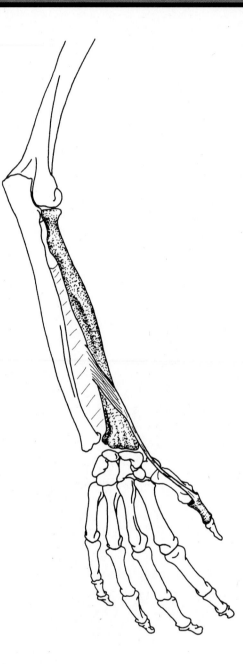

Forearm and hand—dorsal view

■ Origin
Dorsal surface of radius, adjacent part of interosseous membrane

■ Insertion
Base of proximal phalanx of thumb

■ Action
Extends thumb, abducts hand

■ Nerve
Radial nerve (C6, C7)

EXTENSOR POLLICIS LONGUS

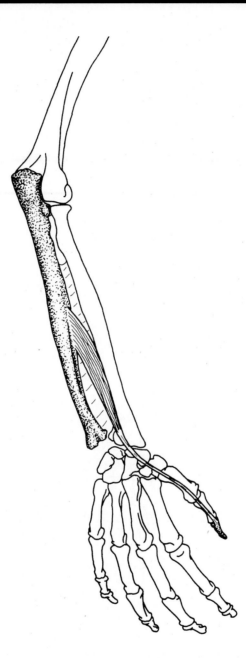

Forearm and hand—dorsal view

■ **Origin**
Middle third of dorsal surface of ulna, interosseous membrane

■ **Insertion**
Base of distal phalanx of thumb

■ **Action**
Extends thumb

■ **Nerve**
Radial nerve (C6–C8)

EXTENSORS OF THE THUMB

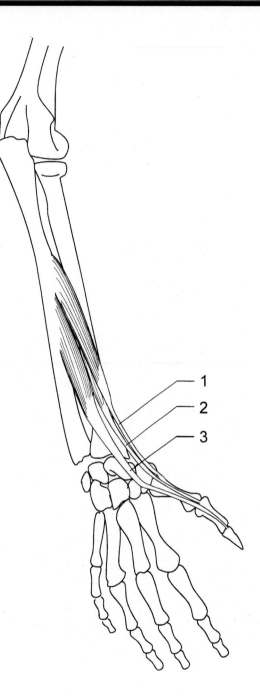

1

2

3

Forearm—posterior view

1. Abductor pollicis longus
2. Extensor pollicis brevis
3. Extensor pollicis longus

EXTENSOR INDICIS

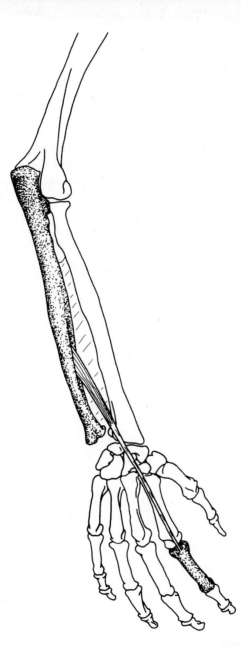

Forearm and hand—dorsal view

■ **Origin**
Posterior surface of ulna and adjacent part of interosseous membrane

■ **Insertion**
Extensor expansion on dorsal surface of proximal phalanx of index finger

■ **Action**
Extends index finger

■ **Nerve**
Radial nerve (C6–C8)

PALMARIS BREVIS

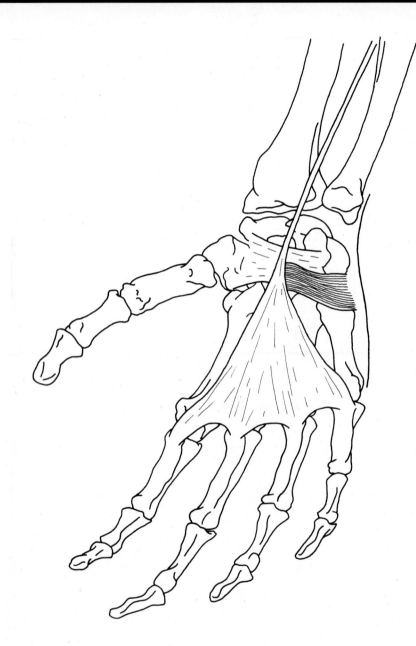

Hand—palmar view

■ **Origin**
Flexor retinaculum, palmar aponeurosis

■ **Insertion**
Skin of the palm

■ **Action**
Corrugates skin of palm

■ **Nerve**
Ulnar nerve (C8)

ABDUCTOR POLLICIS BREVIS *(Thenar Eminence)*

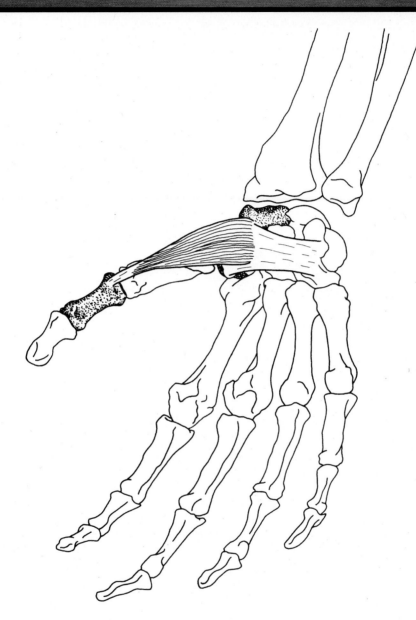

Hand—palmar view

■ Origin
Tubercle of scaphoid, tubercle of trapezium, flexor retinaculum

■ Insertion
Base of proximal phalanx of thumb

■ Action
Abducts thumb and moves it anteriorly, acts together with other muscles of thenar eminence to oppose thumb to other fingers

■ Nerve
Median (C6, C7)

Note: The abductor pollicis brevis, flexor pollicis brevis, and opponens pollicis form the thenar eminence at the base of the thumb.

FLEXOR POLLICIS BREVIS *(Thenar Eminence)*

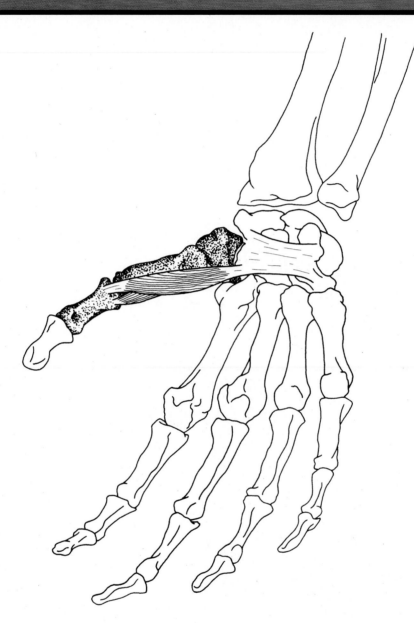

Hand—palmar view

■ **Origin**
Flexor retinaculum and trapezium, and first metacarpal bone

■ **Insertion**
Base of proximal phalanx of thumb

■ **Action**
Flexes metacarpophalangeal joint of thumb, assists in abduction and rotation of thumb, acts together with other muscles of thenar eminence to oppose thumb to other fingers

■ **Nerve**
Lateral portion—median nerve (C6, C7)

Medial portion—ulnar nerve (C8, T1)

OPPONENS POLLICIS *(Thenar Eminence)*

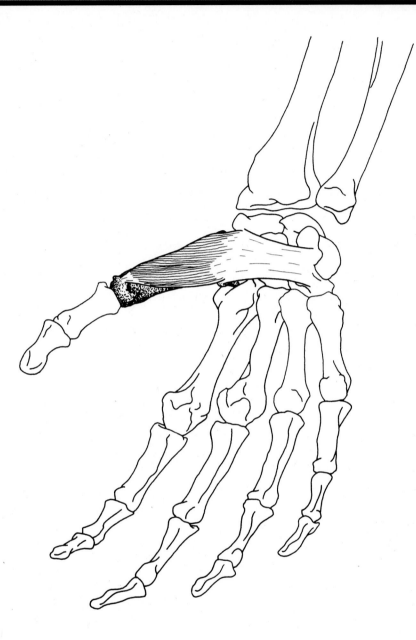

Hand—palmar view

■ **Origin**
Flexor retinaculum, tubercle of trapezium

■ **Insertion**
Lateral border of first metacarpal bone

■ **Action**
Rotates thumb into opposition with fingers, acts together with other muscles of thenar eminence to oppose thumb to other fingers

■ **Nerve**
Median nerve (C6, C7)

ADDUCTOR POLLICIS

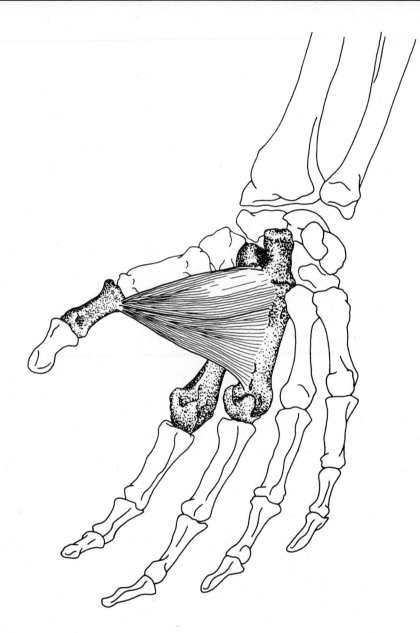

Hand—palmar view

■ Origin
Oblique head—anterior surfaces of second and third metacarpals, capitate, trapezoid

Transverse head—anterior surface of third metacarpal bone

■ Insertion
Medial side of base of proximal phalanx of the thumb

■ Action
Adducts thumb

■ Nerve
Ulnar nerve (C8, T1)

ABDUCTOR DIGITI MINIMI *(Hypothenar Eminence)*

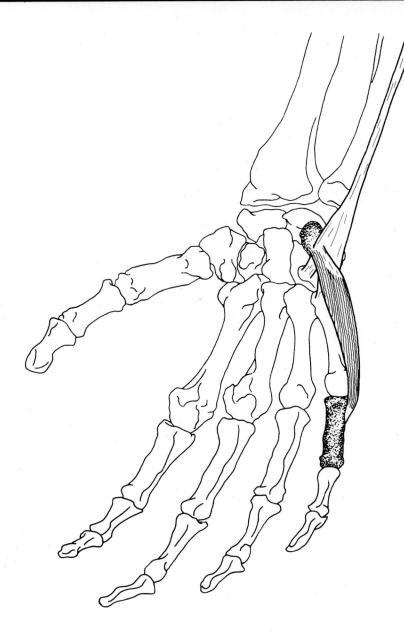

Hand—palmar view

■ Origin
Pisiform bone, tendon of flexor carpi ulnaris

■ Insertion
Medial side of base of proximal phalanx of fifth finger

■ Action
Abducts fifth finger

■ Nerve
Ulnar nerve (C8, T1)

Note: The hypothenar eminence is less prominent than the thenar eminence, and the fifth finger obviously cannot oppose the other digits.

FLEXOR DIGITI MINIMI BREVIS *(Hypothenar Eminence)*

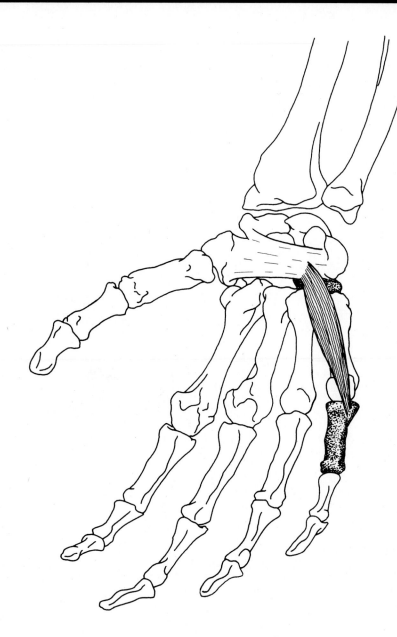

Hand—palmar view

■ Origin
Anterior surface of flexor retinaculum, hook of hamate

■ Insertion
Medial side of base of proximal phalanx of fifth finger

■ Action
Flexes fifth finger at metacarpophalangeal joint

■ Nerve
Ulnar nerve (C8, T1)

OPPONENS DIGITI MINIMI *(Hypothenar Eminence)*

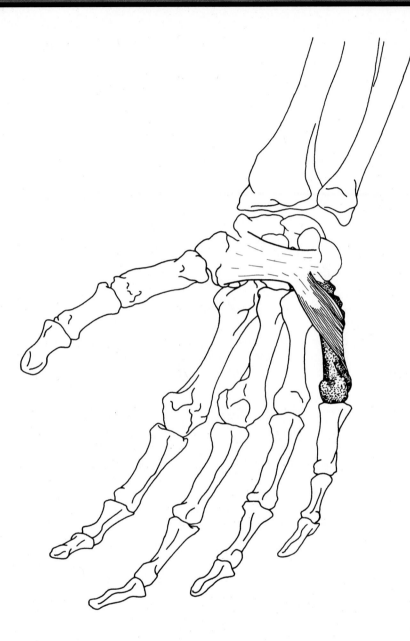

Hand—palmar view

■ Origin
Anterior surface of flexor retinaculum, hook of hamate

■ Insertion
Whole length of medial border of fifth metacarpal bone

■ Action
Rotates fifth metacarpal bone, draws fifth metacarpal bone forward, assists flexor digiti minimi brevis in flexing carpometacarpal joint of fifth finger

■ Nerve
Ulnar nerve (C8, T1)

LUMBRICALES* *(Four Muscles)*

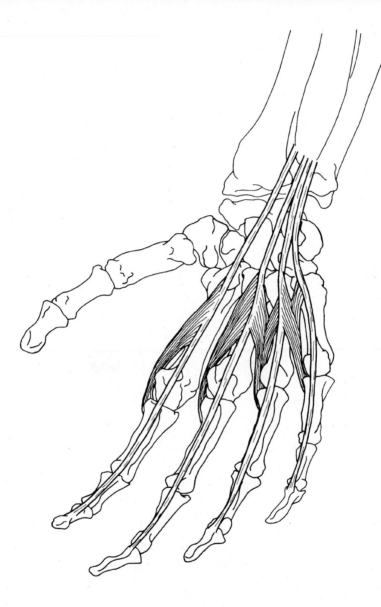

Hand—palmar view

■ Origin
Tendons of flexor digitorum profundus in palm

■ Insertion
Lateral side of corresponding tendon of extensor digitorum on fingers

■ Action
Extend fingers at interphalangeal joints, weakly flex fingers at metacarpophalangeal joints

■ Nerve
Lateral lumbricals (first and second)—median nerve (C6, C7)

Medial lumbricals (third and fourth)—ulnar nerve (C8)

■ Relationships
Assist extensor digitorum communis in extending fingers without hyperextension at the metacarpophalangeal joints

*Associated with the tendons of flexor digitorum profundus.

PALMAR INTEROSSEI

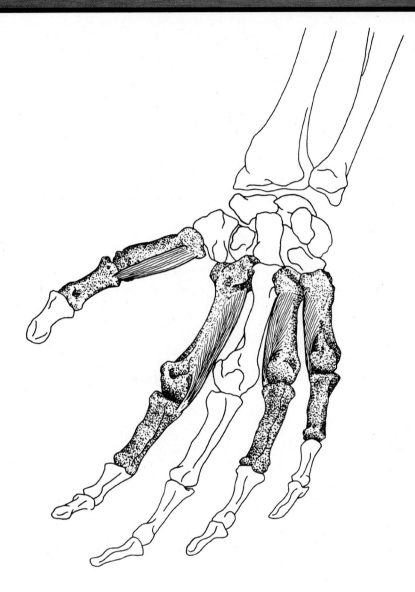

Hand—palmar view

■ Origin

First—medial side of base of first metacarpal bone

Second, third, and fourth—anterior surfaces of second, fourth, and fifth metacarpal bones

■ Insertion

First—medial side of base of proximal phalanx of thumb

Second—medial side of base of proximal phalanx of index finger

Third and fourth—lateral side of proximal phalanges of ring finger and fifth finger

■ Action

Adduct fingers toward center of third finger at metacarpophalangeal joints, assist in flexion of fingers at metacarpophalangeal joints

■ Nerve

Ulnar nerve (C8, T1)

Note: The palmar interosseus of the thumb, called the palmar interosseus of Henle, is usually absent.

DORSAL INTEROSSEI

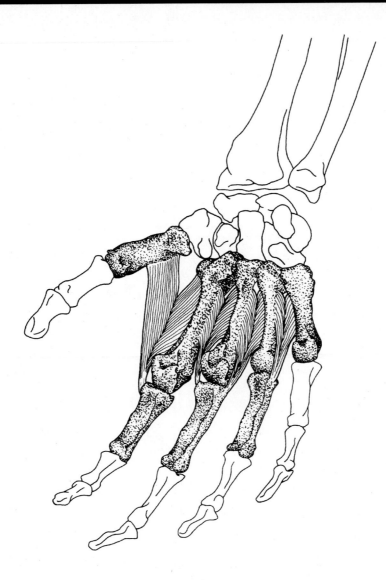

Hand—palmar view

■ Origin
By two heads from adjacent sides of first and second, second and third, third and fourth, and fourth and fifth metacarpal bones

■ Insertion
First—lateral side of base of proximal phalanx of index finger

Second—lateral side of base of proximal phalanx of middle finger

Third—medial side of base of proximal phalanx of middle finger

Fourth—medial side of base of proximal phalanx of ring finger

■ Action
Abduct fingers away from center of third finger at metacarpophalangeal joints, assist in flexion of fingers at metacarpophalangeal joints

■ Nerve
Ulnar nerve (C8, T1)

Muscles of the Hip and Thigh

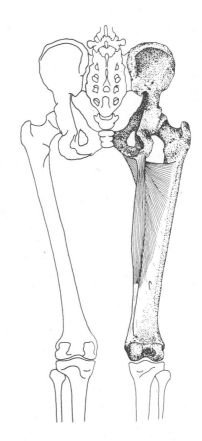

PSOAS MAJOR *(Part of Iliopsoas)*

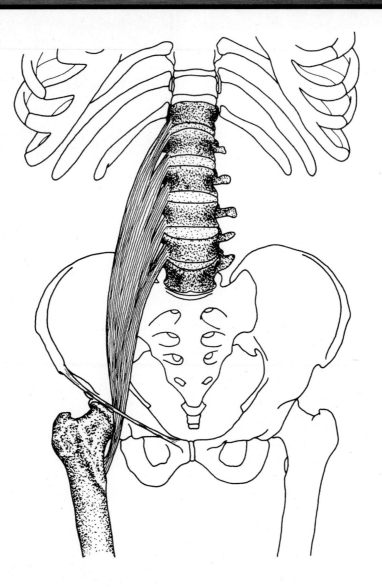

Lumbar region, hip, and thigh—anterior view

■ Origin
Bases of transverse processes of all lumbar vertebrae, bodies of twelfth thoracic and all lumbar vertebrae, intervertebral disks above each lumbar vertebra

■ Insertion
Lesser trochanter of femur

■ Action
Flexes thigh at hip joint, flexes vertebral column*

■ Nerve
Branches from lumbar plexus (L2, L3) and sometimes L1 or L4

Note: Some upper fibers insert onto the hip bone from the arcuate line to the iliopectineal eminence to form the *psoas minor.* This muscle has little function and is frequently absent.

*See note on p 188.

ILIACUS *(Part of Iliopsoas)*

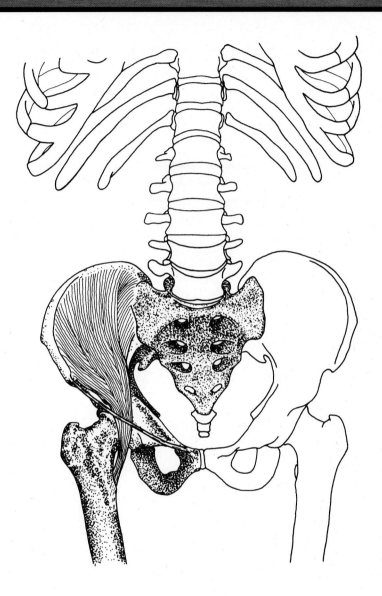

Lumbar region, hip, and thigh—anterior view

■ Origin
Upper two-thirds of iliac fossa, ala of sacrum and adjacent ligaments, anterior inferior iliac spine

■ Insertion
Onto tendon of psoas major, which continues into lesser trochanter of femur (together the two muscles form the iliopsoas)

■ Action
Flexes thigh at hip joint*

■ Nerve
Femoral nerve (L2, L3)

Note: The iliacus brings swinging leg forward in walking or running.

*See note on p 188.

PIRIFORMIS

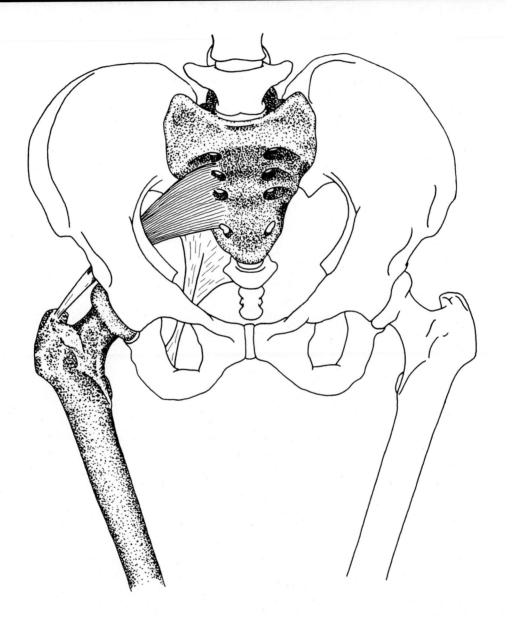

Hip and thigh—anterior view

■ **Origin**
Internal surface of sacrum, sacrotuberous ligament

■ **Insertion**
Upper border of greater trochanter

■ **Action**
Laterally rotates thigh at hip joint, abducts thigh

■ **Nerve**
Anterior rami of first and second sacral nerves

Note: The common peroneal part of the sciatic nerve may emerge through the belly of the piriformis instead of below its inferior border along with the tibial part.

OBTURATOR INTERNUS

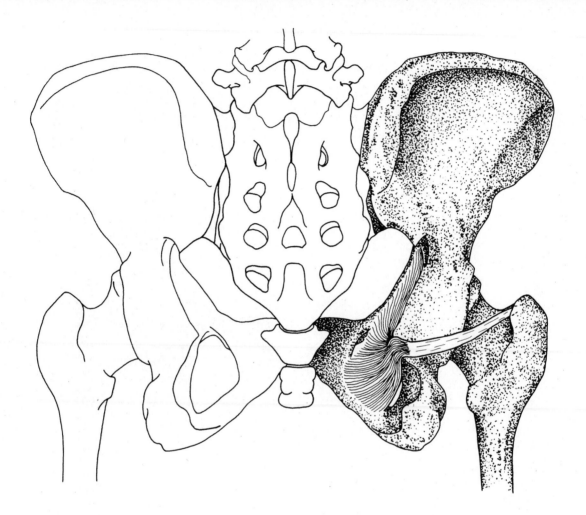

Hip—posterior view

■ **Origin**
Pelvic surface of obturator membrane and surrounding bones (ilium, ischium, pubis)

■ **Insertion**
Common tendon with superior and inferior gemelli to medial surface of upper border of greater trochanter

■ **Action**
Laterally rotates thigh at hip joint

■ **Nerve**
Nerve from sacral plexus (L5, S1–S3)

GEMELLUS SUPERIOR

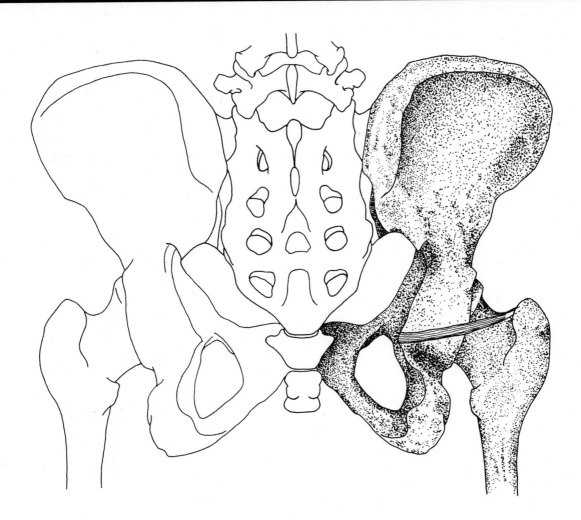

Hip—posterior view

■ Origin
Spine of ischium

■ Insertion
With tendon of obturator internus into medial surface of upper border of greater trochanter

■ Action
Laterally rotates thigh at hip joint

■ Nerve
Branch of nerve to obturator internus from sacral plexus

GEMELLUS INFERIOR

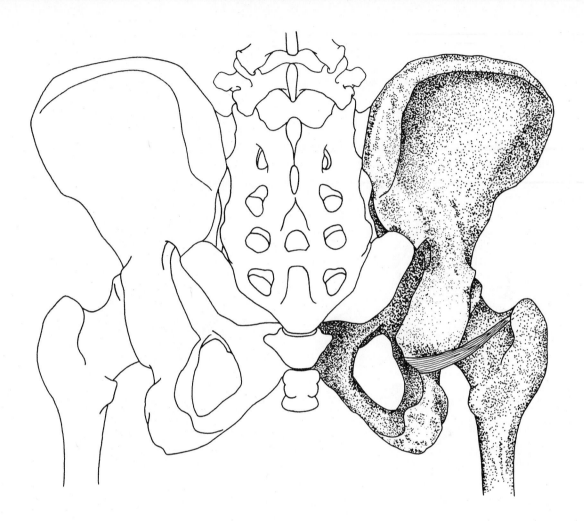

Hip—posterior view

■ **Origin**
Upper margin of ischial tuberosity

■ **Insertion**
With tendon of obturator internus into medial surface
of upper border of greater trochanter

■ **Action**
Laterally rotates thigh at hip joint

■ **Nerve**
Branch of nerve to quadratus femoris from sacral plexus

OBTURATOR EXTERNUS

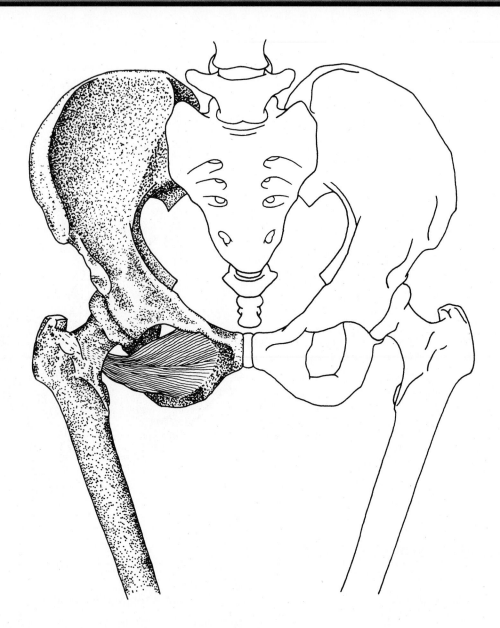

Hip and thigh—anterior view

■ Origin
Outer surface of superior and inferior rami of pubis and ramus of ischium surrounding obturator foramen

■ Insertion
Trochanteric fossa of femur

■ Action
Laterally rotates thigh

■ Nerve
Obturator nerve (L3, L4)

Note: Part of this muscle can be seen posteriorly by separating the gemellus inferior and quadratus femoris. It is deep within this cleft.

QUADRATUS FEMORIS

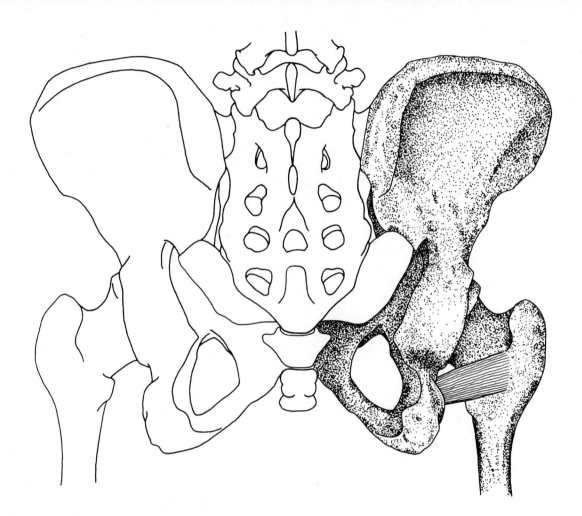

Hip and thigh—posterior view

■ **Origin**
Lateral border of ischial tuberosity

■ **Insertion**
Below intertrochanteric crest (quadrate line)

■ **Action**
Laterally rotates thigh at hip joint

■ **Nerve**
Branch from sacral plexus (L5, S1)

GLUTEUS MAXIMUS

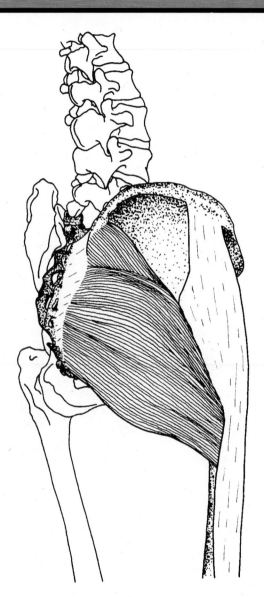

Hip and thigh—lateral view

■ Origin
Outer surface of ilium behind posterior gluteal line, adjacent posterior surface of sacrum and coccyx, sacrotuberous ligament, aponeurosis of erector spinae (sacrospinalis)

■ Insertion
Iliotibial tract of fascia lata, gluteal tuberosity of femur

■ Action
Upper part—abducts, laterally rotates thigh

Lower part—extends, laterally rotates thigh, extends trunk, assists in adduction of thigh

■ Nerve
Inferior gluteal nerve (L5, S1, S2)

Note: This is not a postural muscle; it is not used in walking but only in forceful extension, as in running, climbing, or rising from a seated position.

GLUTEUS MEDIUS

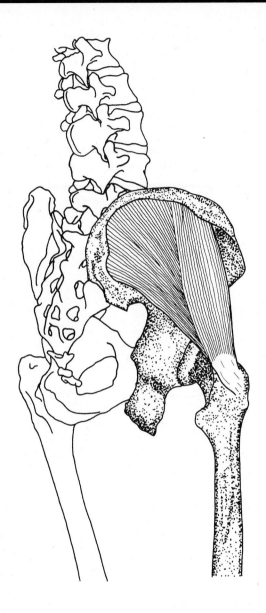

Hip and thigh—lateral view

■ Origin
Outer surface of ilium inferior to iliac crest

■ Insertion
Lateral surface of greater trochanter

■ Action
Abducts femur at hip joint and rotates thigh medially

■ Nerve
Superior gluteal nerve (L4, L5, S1)

Note: In locomotion, this muscle (along with the gluteus minimus) prevents the pelvis from dropping (adduction of thigh) toward the opposite swinging leg.

GLUTEUS MINIMUS

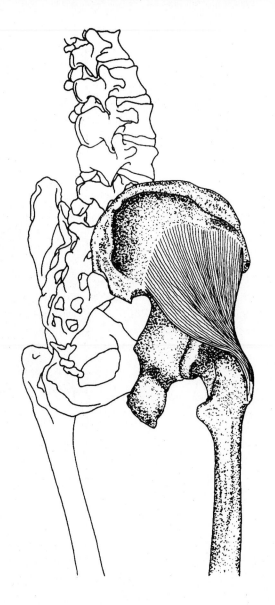

Hip and thigh—lateral view

■ Origin
Outer surface of ilium between middle (anterior) and inferior gluteal lines

■ Insertion
Anterior surface of greater trochanter

■ Action
Abducts femur at hip joint and rotates thigh medially

■ Nerve
Superior gluteal nerve (L4, L5, S1)

See note on p 173.

MUSCLES OF THE HIP

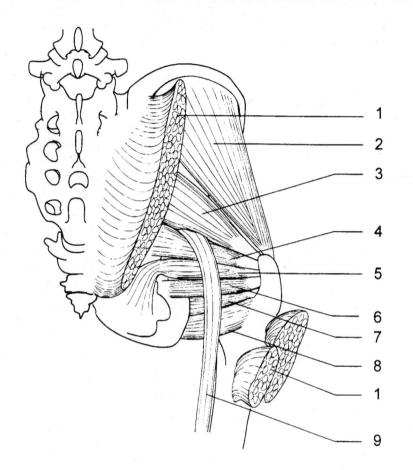

1
2
3
4
5
6
7
8
1
9

Hip—posterior view

1. Gluteus maximus (cut)
2. Gluteus medius
3. Piriformis
4. Gemellus superior
5. Obturator internus

6. Gemellus inferior
7. Obturator externus
8. Quadratus femoris
9. Sciatic nerve

Note: Gemellus inferior and quadratus femoris have been shown separated to expose the deeply placed obturator externus.

TENSOR FASCIAE LATAE

Hip and thigh—lateral view

■ Origin
Outer edge of iliac crest between anterior superior iliac spine and iliac tubercle

■ Insertion
Iliotibial tract on upper part of thigh

■ Action
Flexes, abducts thigh

■ Nerve
Superior gluteal nerve (L4, L5, S1)

Note: This muscle, along with gluteus maximus, draws the fascia lata upward, stabilizing the knee.

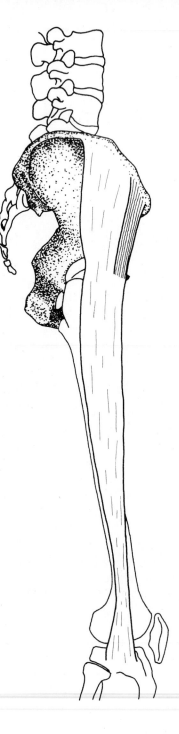

SARTORIUS

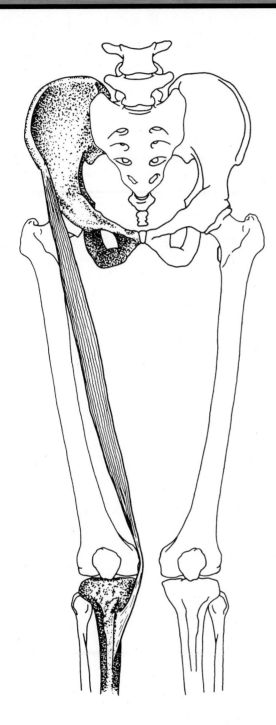

Hip, thigh, and leg—anterior view

■ Origin
Anterior superior iliac spine and area immediately below it

■ Insertion
Upper part of medial surface of shaft of tibia

■ Action
Flexes, abducts, and laterally rotates thigh at hip joint, flexes and slightly medially rotates leg at knee joint after flexion

■ Nerve
Femoral nerve (L2, L3)

■ Relationships
Insertions of sartorius, gracilis, and semitendinosus fuse on the medial tibia; these tendons, called the pes anserinus (goose foot), give medial support to the knee

Note: This muscle is used to bring swinging leg forward in walking and running.

RECTUS FEMORIS *(One of quadriceps femoris)*

Hip, thigh, and leg—anterior view

■ Origin
Anterior head—anterior inferior iliac spine

Posterior head—ilium above acetabulum

■ Insertion
Patella, then by patellar ligament to tuberosity of the tibia

■ Action
Extends leg at knee joint, flexes thigh at hip joint

■ Nerve
Femoral nerve (L2–L4)

Note: This muscle is used when thigh flexion and leg extension are needed together, such as in kicking a football. In walking, the quadriceps prevent the knee from flexing during heel strike and early support phase.

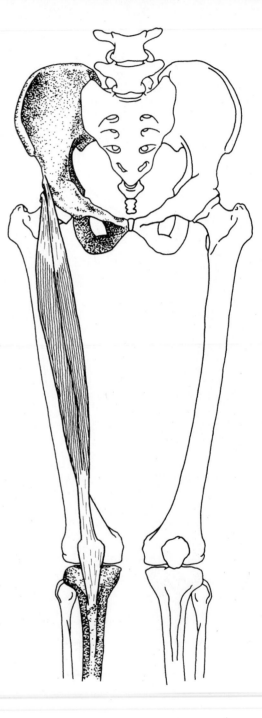

VASTUS LATERALIS *(One of quadriceps femoris)*

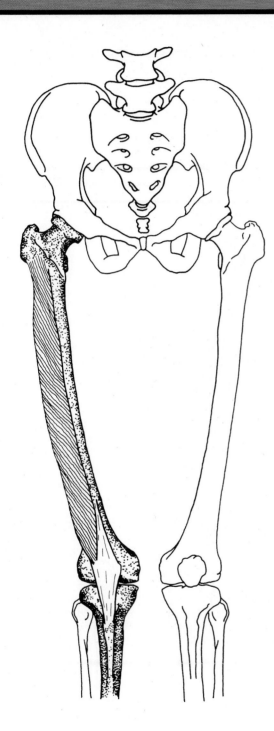

Hip, thigh, and leg—anterior view

■ Origin
Intertrochanteric line, inferior border of greater trochanter, gluteal tuberosity, lateral lip of linea aspera of femur

■ Insertion
Lateral margin of patella, then by patellar ligament to tuberosity of tibia

■ Action
Extends leg at knee joint

■ Nerve
Femoral nerve (L2–L4)

VASTUS MEDIALIS *(One of quadriceps femoris)*

Hip, thigh, and leg—anterior view

■ Origin
Intertrochanteric line, medial lip of linea aspera of femur, medial intermuscular septum, medial supracondylar ridge

■ Insertion
Medial border of the patella, then by patellar ligament into tibial tuberosity, medial condyle of tibia

■ Action
Extends leg at knee joint

■ Nerve
Femoral nerve (L2–L4)

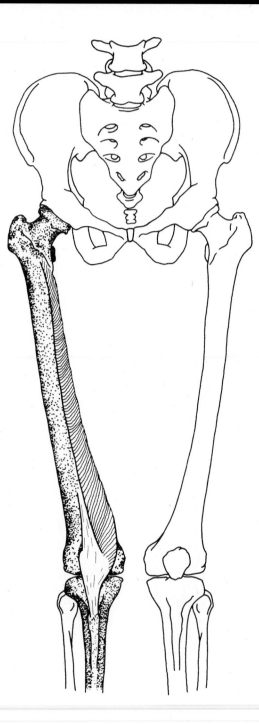

VASTUS INTERMEDIUS *(One of quadriceps femoris)*

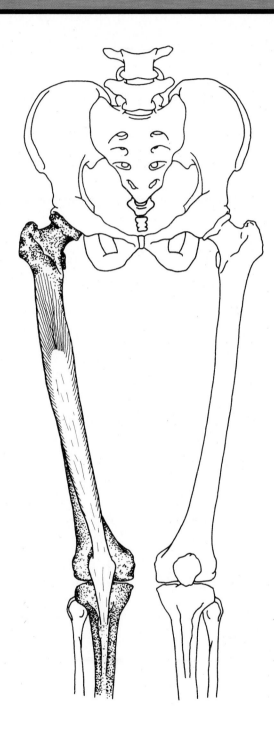

Hip, thigh, and leg—anterior view

■ Origin
Anterior and lateral surfaces of upper two-thirds of femur, lateral intermuscular septum, linea aspera, lateral supracondylar ridge

■ Insertion
Deep aspect of quadriceps tendon, then through patella to tibial tuberosity

■ Action
Extends leg at knee joint

■ Nerve
Femoral nerve (L2–L4)

Note: A few bundles of fibers from this muscle insert onto the upper part of the joint capsule of the knee. They probably draw the capsule superiorly during extension of the leg, preventing it from binding in the joint. They are called *articularis genus.*

MUSCLES OF THE ANTERIOR THIGH

Hip and thigh—anterior view

1. Tensor fasciae latae
2. Iliotibial tract
3. Vastus lateralis (quadriceps femoris)
4. Vastus intermedius (quadriceps femoris)
5. Rectus femoris (cut) (quadriceps femoris)
6. Sartorius
7. Vastus medialis (quadriceps femoris)

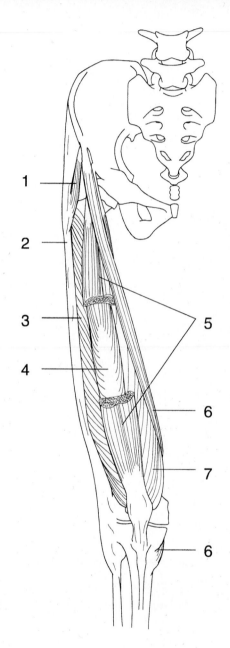

BICEPS FEMORIS *(Part of Hamstrings)*

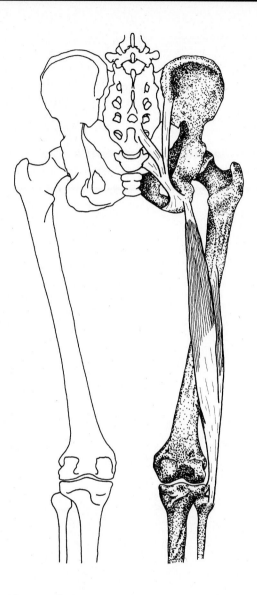

Hip and thigh—posterior view

■ Origin
Long head—ischial tuberosity, sacrotuberous ligament

Short head—linea aspera, lateral supracondylar ridge, lateral intermuscular septum

■ Insertion
Lateral side of head of fibula and lateral condyle of tibia

■ Action
Flexes leg at knee joint, long head also extends thigh at hip joint

■ Nerve
Long head—tibial part of sciatic nerve (S1–S3)

Short head—common peroneal part of sciatic nerve (L5, S1, S2)

Note: During walking or running, the hamstrings are used to slow down the leg at the end of its swing and prevent the trunk from flexing at the hip. They are susceptible to being strained by resisting the momentum of these body parts.

SEMITENDINOSUS *(Part of Hamstrings)*

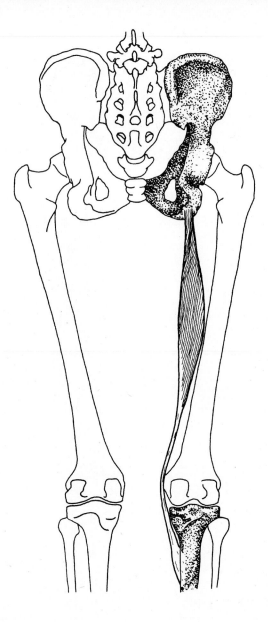

Hip and thigh—posterior view

■ **Origin**
Ischial tuberosity

■ **Insertion**
Medial surface of shaft of tibia

■ **Action**
Flexes and slightly medially rotates leg at knee joint
after flexion, extends thigh at hip joint

■ **Nerve**
Tibial portion of sciatic nerve (L5, S1, S2)

See note on p 183 and relationships section on p 177.

SEMIMEMBRANOSUS *(Part of Hamstrings)*

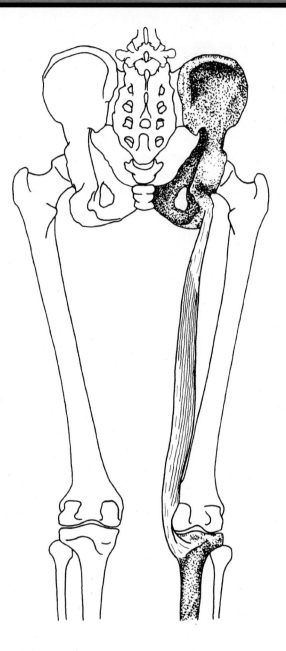

Hip and thigh—posterior view

■ Origin
Ischial tuberosity

■ Insertion
Posterior part of medial condyle of tibia

■ Action
Flexes and slightly medially rotates leg at knee joint after flexion, extends thigh at hip joint

■ Nerve
Tibial portion of sciatic nerve (L5, S1, S2)

See note on p 183.

HAMSTRING MUSCLES

Hip and thigh—posterior view

1. Sciatic nerve
2. Quadratus femoris
3. Biceps femoris
4. Semimembranosus
5. Semitendinosus
6. Tibial nerve
7. Common peroneal nerve

Note: The common peroneal nerve is exposed to compression and damage as it passes over the head of the fibula. The quadratus femoris, a lateral rotator, is included for reference.

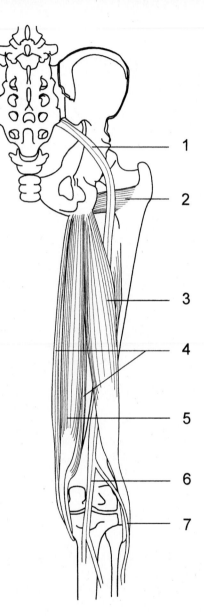

GRACILIS

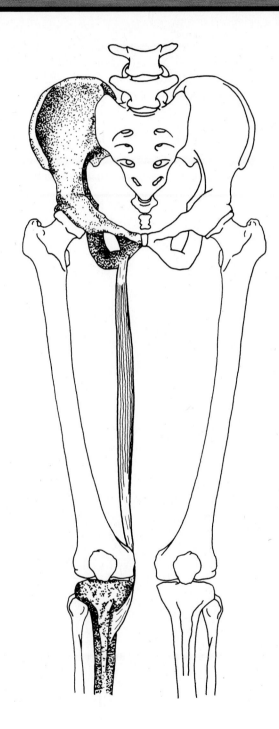

Hip and thigh—anterior view

■ Origin
Lower margin of body and inferior ramus of pubis

■ Insertion
Upper part of medial surface of shaft of tibia

■ Action
Adducts thigh at hip joint and flexes leg at knee joint; with leg flexed, assists in medial rotation

■ Nerve
Obturator nerve (L3, L4)

See relationships section on p 177.

PECTINEUS

Hip and thigh—anterior view

■ Origin
Pectineal line on superior ramus of pubis

■ Insertion
From lesser trochanter to linea aspera of femur

■ Action
Flexes thigh, assists in adduction when hip is flexed

■ Nerve
Femoral nerve (L2–L4), (sometimes a branch of obturator nerve)

Note: The rotational component of thigh muscle action depends upon the starting position of the hip joint. The pectineus, adductor longus, adductor brevis, and psoas major probably assist in medial rotation when the thigh is in anatomical position but may shift to assisting in lateral rotation as the thigh flexes and abducts.

The iliacus and adductor tubercle part of adductor magnus probably assist in medial rotation throughout the range of motion of the hip joint while the linea aspera part of the adductor magnus may be a slight lateral rotator.

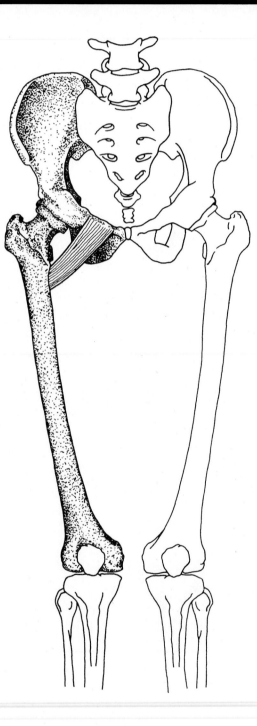

ADDUCTOR LONGUS

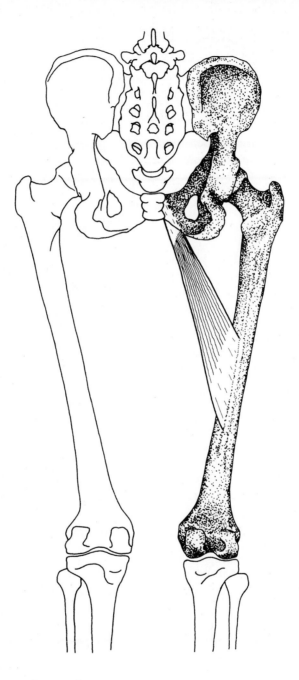

Hip and thigh—posterior view

■ Origin
Anterior of body of pubis

■ Insertion
Medial lip of linea aspera

■ Action
Adducts, flexes thigh, assists in medial rotation*

■ Nerve
Obturator nerve (L3, L4)

*See note on p 188.

ADDUCTOR BREVIS

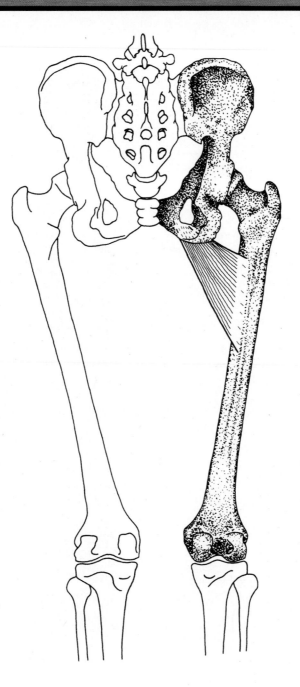

Hip and thigh—posterior view

■ Origin
Outer surface of inferior ramus of pubis

■ Insertion
From below lesser trochanter to linea aspera and into proximal part of linea aspera

■ Action
Adducts thigh, assists in flexion, medial rotation*

■ Nerve
Obturator nerve (L3, L4)

*See note on p 188.

ADDUCTOR MAGNUS

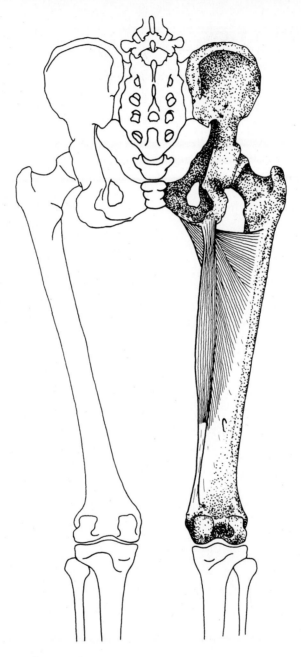

Hip and thigh—posterior view

■ Origin
Inferior ramus of pubis, and ramus and lower part of tuberosity of ischium

■ Insertion
Linea aspera, adductor tubercle of femur

■ Action
Adducts, extends thigh, lower portion (adductor tubercle insertion) assists in medial rotation*

■ Nerve
Obturator nerve (L3, L4), sciatic nerve

*See note on p 188.

HIP FLEXORS AND ADDUCTORS

Hip and thigh—anterior view

1. Psoas major
2. Iliacus
3. Inguinal ligament
4. Femoral nerve, vein, artery
5. Pectineus
6. Adductor brevis
7. Adductor longus (cut)
8. Adductor magnus
9. Gracilis

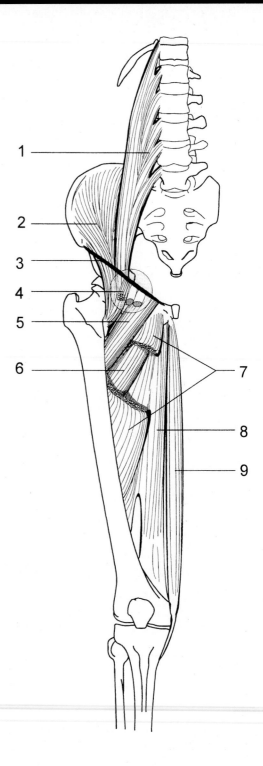

CHAPTER NINE

Muscles of the
Leg and Foot

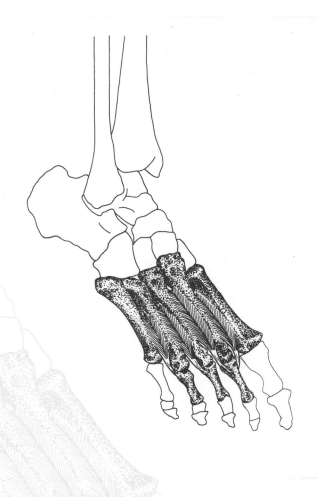

TIBIALIS ANTERIOR

Leg—anterolateral view

■ Origin
Lateral condyle of tibia, upper half of lateral surface of tibia, interosseous membrane

■ Insertion
Medial side and plantar surface of medial cuneiform bone, and base of first metatarsal bone

■ Action
Dorsiflexes foot at ankle joint, inverts (supinates) foot

■ Nerve
Deep peroneal nerve (L4, L5, S1)

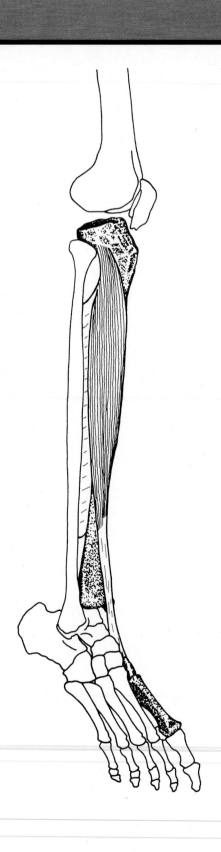

EXTENSOR HALLUCIS LONGUS

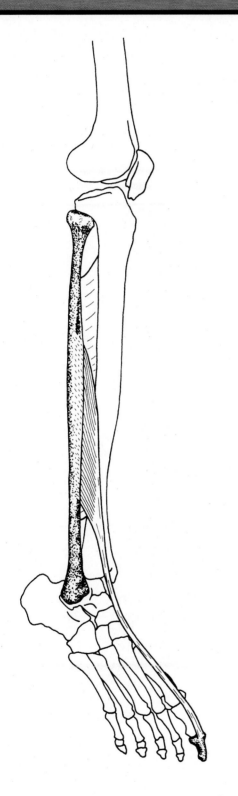

Leg—anterolateral view

■ Origin
Middle half of anterior surface of fibula and interosseous membrane

■ Insertion
Base of distal phalanx of great toe

■ Action
Extends, hyperextends great toe, dorsiflexes and inverts (supinates) foot

■ Nerve
Deep peroneal nerve (L4, L5, S1)

EXTENSOR DIGITORUM LONGUS

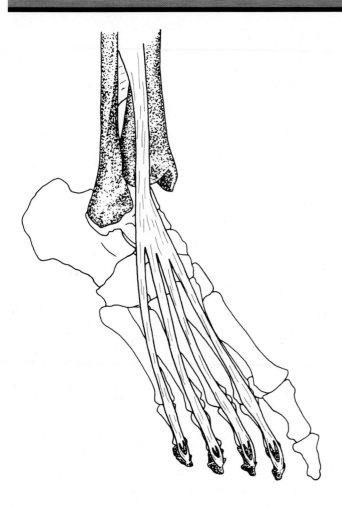

Foot—anterolateral view

■ Origin
Upper two-thirds of anterior surface of fibula, interosseous membrane, lateral condyle of tibia

■ Insertion
Along dorsal surface of four lateral toes, and then to bases of middle and distal phalanges

■ Action
Extends toes, dorsiflexes foot at ankle, everts foot

■ Nerve
Deep peroneal nerve (L4, L5, S1)

Note: The lower lateral part of this muscle makes a separate insertion onto the dorsal surface of the fifth metatarsal and is called *peroneus tertius.*

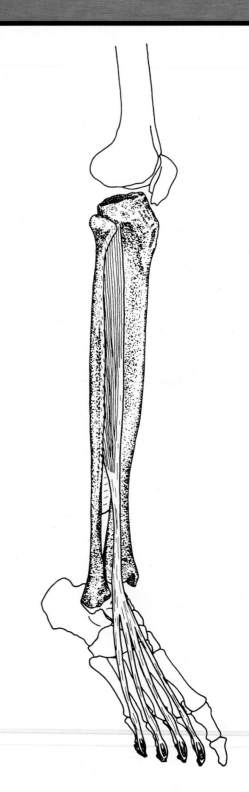

Leg—anterolateral view

PERONEUS TERTIUS *(Lower lateral part of extensor digitorum longus)*

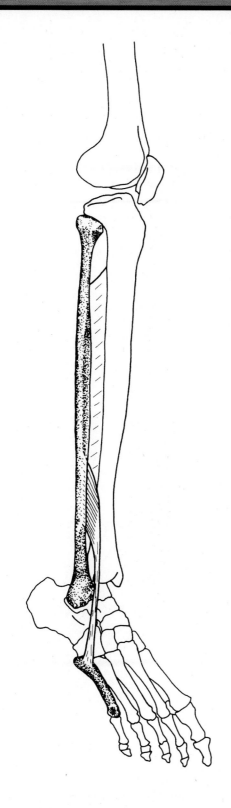

Leg—anterolateral view

■ Origin
Lower third of anterior surface of fibula and interosseous membrane

■ Insertion
Dorsal surface of base of fifth metatarsal bone

■ Action
Dorsiflexes and everts foot

■ Nerve
Deep peroneal nerve (L4, L5, S1)

ANTERIOR AND LATERAL LEG MUSCLES

Leg—anterolateral view

1. Peroneus longus
2. Peroneus brevis
3. Peroneus tertius
4. Tibialis anterior
5. Extensor retinaculum
6. Extensor hallucis longus
7. Extensor digitorum longus
8. Extensor digitorum brevis

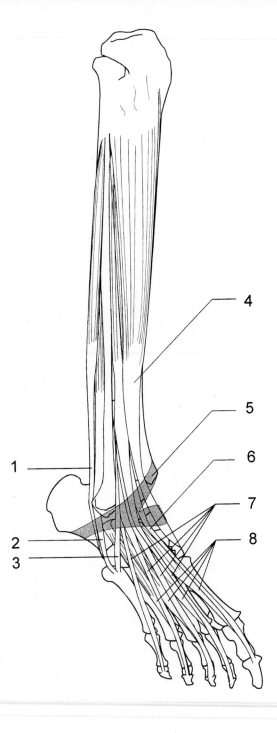

GASTROCNEMIUS *(Part of triceps surae)*

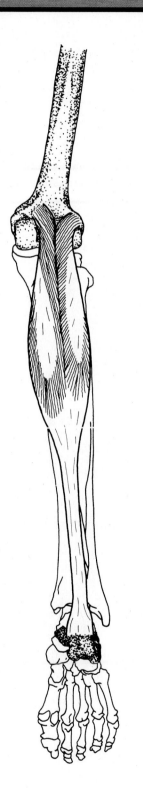

Leg—posterior view

■ Origin
Lateral head—lateral condyle and posterior surface of femur

Medial head—popliteal surface of femur above medial condyle

■ Insertion
Posterior surface of the calcaneus

■ Action
Plantar flexes foot, flexes leg at knee

■ Nerve
Tibial nerve (S1, S2)

SOLEUS *(Part of triceps surae)*

Leg—posterior view

■ Origin
Posterior surface of the tibia (soleal line), upper third of posterior surface of fibula, fibrous arch between tibia and fibula

■ Insertion
Posterior surface of the calcaneus

■ Action
Plantar flexes foot

■ Nerve
Tibial nerve (S1, S2)

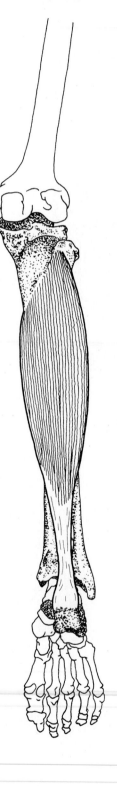

PLANTARIS

Leg—posterior view

■ Origin
Lateral supracondylar ridge of femur, oblique popliteal ligament

■ Insertion
Posterior surface of the calcaneus

■ Action
Plantar flexes foot, flexes leg

■ Nerve
Tibial nerve (L4, L5, S1)

MUSCLES OF THE CALF

Leg—posterior view

1. Soleus
2. Plantaris
3. Gastrocnemius (cut)

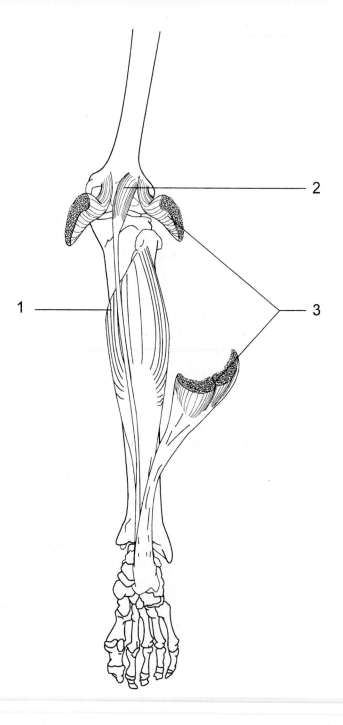

POPLITEUS

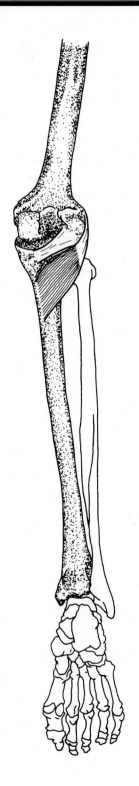

Leg—posterior view

■ Origin
Lateral surface of lateral condyle of femur

■ Insertion
Upper part of posterior surface of tibia

■ Action
Rotates leg medially, flexes leg

■ Nerve
Tibial nerve (L4, L5, S1)

Note: Stern contends that this muscle stabilizes the knee by preventing lateral rotation of the tibia during medial rotation of the thigh while the foot is planted.

Reference: Stern, J. T. *Essentials of Gross Anatomy,* F. A. Davis Company, Philadelphia, 1988.

FLEXOR HALLUCIS LONGUS

Leg—posterior view

■ Origin
Lower two-thirds of posterior surface of shaft of fibula, posterior intermuscular septum, interosseous membrane

■ Insertion
Base of distal phalanx of great toe

■ Action
Flexes distal phalanx of great toe, assists in plantar flexing foot, inverts foot

■ Nerve
Tibial nerve (L5, S1, S2)

Note: This muscle is important in pushing off the surface in walking, running, jumping.

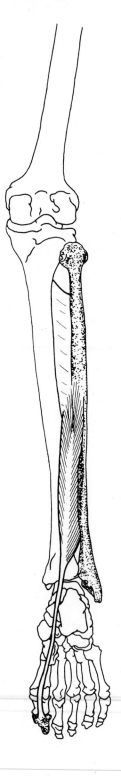

FLEXOR DIGITORUM LONGUS

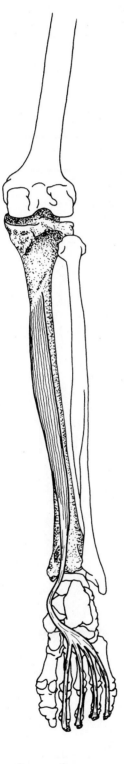

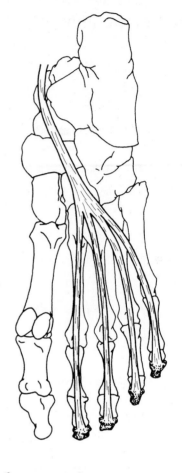

Foot—plantar view

■ Origin
Medial part of posterior surface of tibia

■ Insertion
Bases of distal phalanges of second, third, fourth, and fifth toes

■ Action
Flexes distal phalanges of lateral four toes, assists in plantar flexing foot, inverts foot

■ Nerve
Tibial nerve (L5, S1)

Leg—posterior view

TIBIALIS POSTERIOR

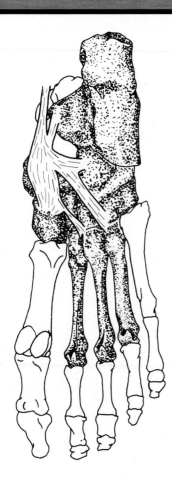

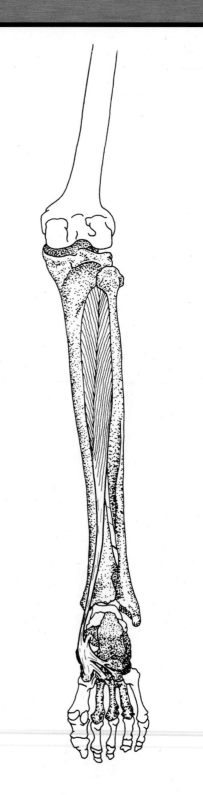

Foot—plantar view

■ Origin
Lateral part of posterior surface of tibia, interosseous membrane, proximal half of posterior surface of fibula

■ Insertion
Tuberosity of navicular bone, cuboid, cuneiforms, second, third, and fourth metatarsals, sustentaculum tali of calcaneus

■ Action
Plantar flexes, inverts foot

■ Nerve
Tibial nerve (L5, S1)

Leg—posterior view

DEEP POSTERIOR LEG MUSCLES

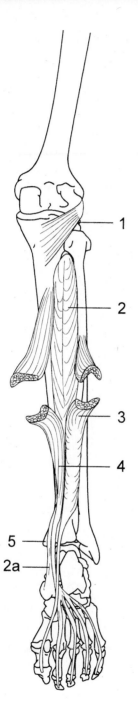

Leg—posterior view

1. Popliteus
2. Tibialis posterior
2a. Tendon of tibialis posterior
3. Flexor hallucis longus (cut)
4. Flexor digitorum longus (cut)
5. Medial malleolus

PERONEUS LONGUS

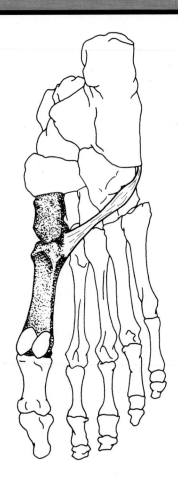

Foot—plantar view

■ Origin
Upper two-thirds of lateral surface of fibula

■ Insertion
Lateral side of medial cuneiform, base of first metatarsal

■ Action
Plantar flexes, everts foot

■ Nerve
Superficial peroneal nerve (L4, L5, S1)

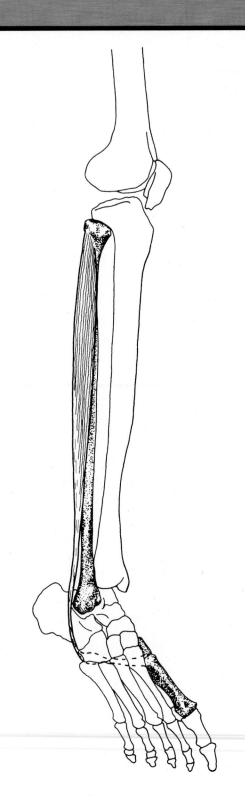

Leg—anterolateral view

PERONEUS BREVIS

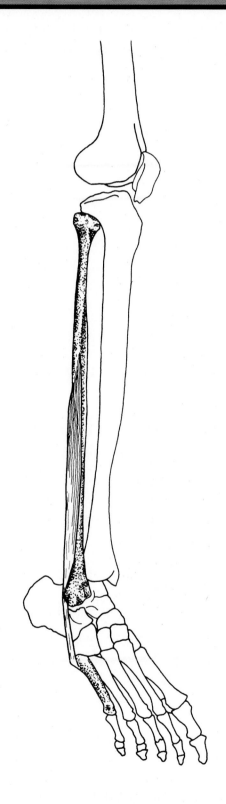

Leg—anterolateral view

■ Origin
Lower two-thirds of lateral surface of fibula

■ Insertion
Lateral side of base of fifth metatarsal bone

■ Action
Everts, plantar flexes foot

■ Nerve
Superficial peroneal nerve (L4, L5, S1)

EXTENSOR DIGITORUM BREVIS

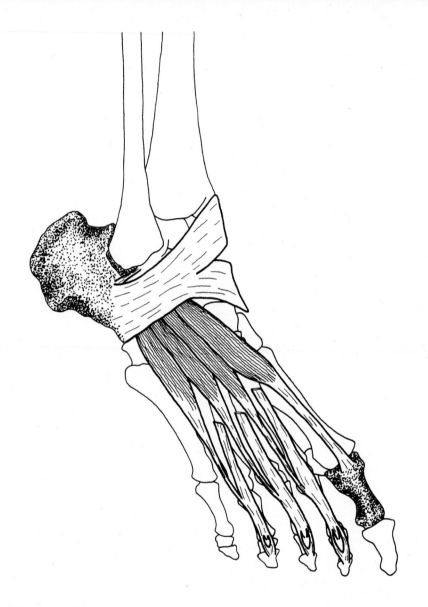

Foot—anterolateral view

■ Origin
Anterior and lateral surfaces of calcaneus, lateral talocalcaneal ligament, inferior extensor retinaculum

■ Insertion
Into base of proximal phalanx of great toe; into lateral sides of tendons of extensor digitorum longus of second, third, and fourth toes

■ Action
Extends the four toes

■ Nerve
Deep peroneal nerve (L5, S1)

ABDUCTOR HALLUCIS *(First Layer)*

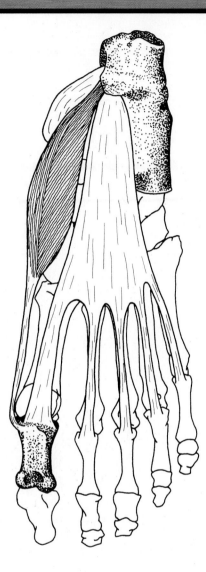

Foot—plantar view

■ **Origin**
Tuberosity of calcaneus, flexor retinaculum, plantar aponeurosis

■ **Insertion**
Medial side of base of proximal phalanx of great toe

■ **Action**
Stabilizes great toe (with adductor hallucis)

■ **Nerve**
Medial plantar nerve (L4, L5)

Note: The muscles of the sole of the foot can be divided into four layers (from superficial to deep):

First layer—abductor hallucis, flexor digitorum brevis, abductor digiti minimi

Second layer—quadratus plantae, lumbricales (tendons of flexor hallucis longus and flexor digitorum longus pass through this layer)

Third layer—flexor hallucis brevis, adductor hallucis, flexor digiti minimi brevis

Fourth layer—interossei (tendons of tibialis posterior and peroneus longus pass through this layer)

FLEXOR DIGITORUM BREVIS *(First Layer)*

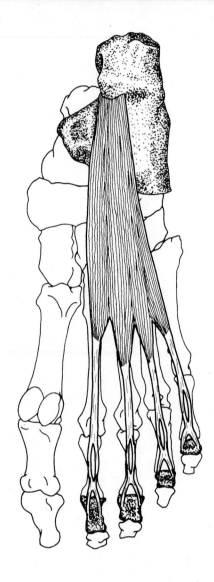

Foot—plantar view

■ **Origin**
Tuberosity of calcaneus, plantar aponeurosis

■ **Insertion**
Sides of middle phalanges of second to fifth toes

■ **Action**
Flexes proximal phalanges and extends distal phalanges of second through fifth toes

■ **Nerve**
Medial plantar nerve (L4, L5)

ABDUCTOR DIGITI MINIMI *(First Layer)*

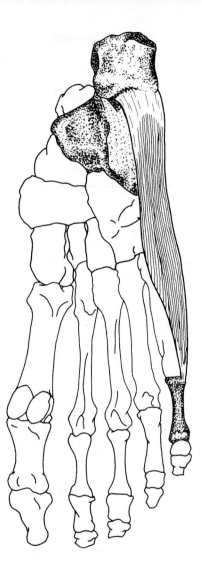

Foot—plantar view

■ Origin
Tuberosity of calcaneus, plantar aponeurosis

■ Insertion
Lateral side of proximal phalanx of fifth toe

■ Action
Abducts fifth toe

■ Nerve
Lateral plantar nerve (S1, S2)

QUADRATUS PLANTAE *(Second Layer)*

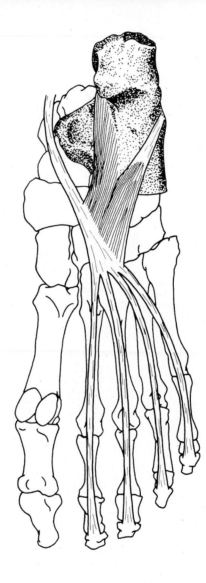

Foot—plantar view

■ **Origin**

Medial head—medial surface of calcaneus

Lateral head—lateral border of inferior surface of calcaneus

■ **Insertion**

Lateral margin of tendon of flexor digitorum longus

■ **Action**

Flexes terminal phalanges of second through fifth toes

■ **Nerve**

Lateral plantar nerve (S1, S2)

LUMBRICALES *(Second Layer)*

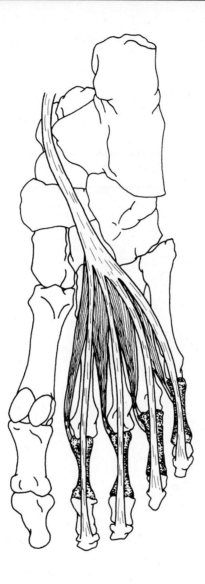

Foot—plantar view

■ Origin
Tendons of flexor digitorum longus

■ Insertion
Dorsal surfaces of proximal phalanges

■ Action
Flex proximal phalanges of second through fifth toes

■ Nerve
First lumbricalis—medial plantar nerve (L4, L5)

Second through fifth lumbricales—lateral plantar nerve (S1, S2)

FLEXOR HALLUCIS BREVIS *(Third Layer)*

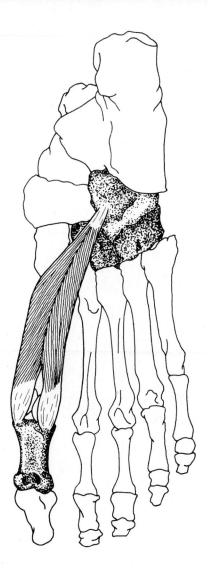

Foot—plantar view

■ **Origin**
Cuboid bone, lateral cuneiform bone

■ **Insertion**
Medial part—medial side of base of proximal phalanx of great toe

Lateral part—lateral side of base of proximal phalanx of great toe

■ **Action**
Flexes proximal phalanx of great toe

■ **Nerve**
Medial plantar nerve (L4, L5, S1)

ADDUCTOR HALLUCIS *(Third Layer)*

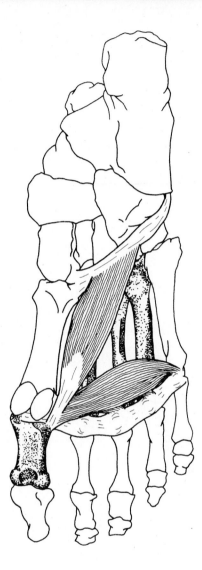

Foot—plantar view

■ Origin
Oblique head—second, third, and fourth metatarsal bones, and sheath of peroneus longus tendon

Transverse head—plantar metatarsophalangeal ligaments of third, fourth, and fifth toes, and transverse metatarsal ligaments

■ Insertion
Lateral side of base of proximal phalanx of great toe

■ Action
Stabilizes great toe (with abductor hallucis)

■ Nerve
Lateral plantar nerve (S1, S2)

FLEXOR DIGITI MINIMI BREVIS *(Third Layer)*

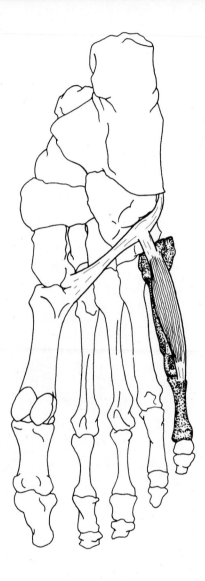

Foot—plantar view

■ Origin
Base of fifth metatarsal, sheath of peroneus longus tendon

■ Insertion
Lateral side of base of proximal phalanx of fifth toe

■ Action
Flexes proximal phalanx of fifth toe

■ Nerve
Lateral plantar nerve (S1, S2)

DORSAL INTEROSSEI *(Fourth Layer; Four Muscles)*

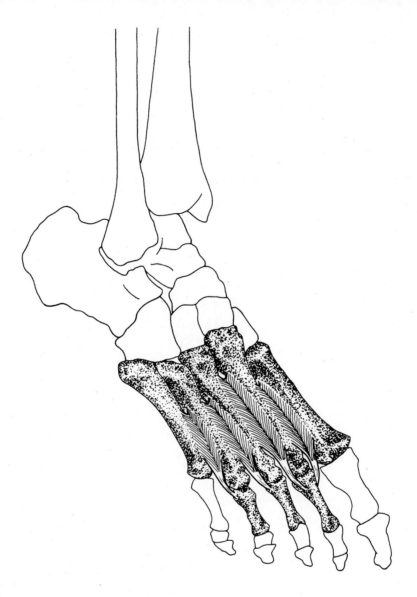

Foot—anterolateral view

■ Origin
Adjacent sides of metatarsal bones

■ Insertion
Bases of proximal phalanges

First—medial side of proximal phalanx of second toe

Second, third, fourth—lateral sides of proximal phalanges of second, third, and fourth toes

■ Action
Abduct toes, flex proximal phalanges

■ Nerve
Lateral plantar nerve (S1, S2)

PLANTAR INTEROSSEI *(Fourth Layer; Three Muscles)*

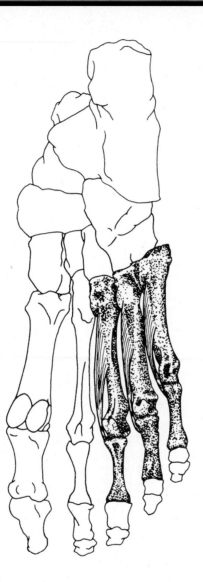

Foot—plantar view

■ Origin
Bases and medial sides of third, fourth, and fifth metatarsal bones

■ Insertion
Medial sides of bases of proximal phalanges of same toes

■ Action
Adduct toes, flex proximal phalanges

■ Nerve
Lateral plantar nerve (S1, S2)

Alphabetical Listing of Muscles

Index

Notes

Notes

Notes

Notes